571.6

AS/A-LEVEL YEAR 1

STUDENT GUIDE

OCR

Biology A

Modules 3 and 4

Exchange and transport

Biodiversity, evolution and disease

Richard Fosbery

PHILIP ALLAN FOR
HODDER
EDUCATION
AN HACHETTE UK COMPANY

Philip Allan, an imprint of Hodder Education, an Hachette UK company, Blenheim Court, George Street, Banbury, Oxfordshire OX16 5BH

Orders

Bookpoint Ltd, 130 Park Drive, Milton Park, Abingdon, Oxfordshire OX14 4SE

tel: 01235 827827

fax: 01235 400401

e-mail: education@bookpoint.co.uk

Lines are open 9.00 a.m.–5.00 p.m., Monday to Saturday, with a 24-hour message answering service. You can also order through the Hodder Education website: www.hoddereducation. co.uk

This guide has been written specifically to support students preparing for the OCR AS and A-level Biology examinations. The content has been neither approved nor endorsed by OCR and remains the sole responsibility of the author.

Cover photo: Argonautis/Fotolia; p. 33, Ian Couchman/CIE Laboratory, Cambridge; p. 37, John Luttick and Lawrence Wesson, James Allen's Girls' School; p. 38 left, Margot Fosbery; p. 38 right, Richard Fosbery; p. 52, Dr Clare van der Willigen; p. 63 top; Auscape International Pty Ltd/Alamy Stock Photo; p. 63 bottom, Robert Hoetink/Fotolia

Typeset by Integra Software Services Pvt. Ltd, Pondicherry, India

Printed in Italy

Hachette UK's policy is to use papers that are natural, renewable and recyclable products and made from wood grown in sustainable forests. The logging and manufacturing processes are expected to conform to the environmental regulations of the country of origin.

Contents

Content Guidance

Questions & Answers

∎ Getting the most from this book

Exam-style questions

Commentary on the questions

Tips on what you need to do to gain full marks, indicated by the icon **e**

Sample student answers

Practise the questions, then look at the student answers that follow.

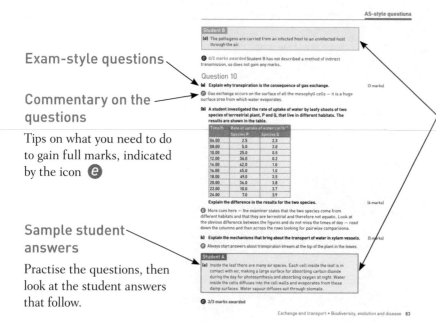

Commentary on sample student answers

Find out how many marks each answer would be awarded in the exam and then read the comments (preceded by the icon **e**) following each student answer showing exactly how and where marks are gained or lost.

■ About this book

This guide is the second in a series of four covering the OCR AS and A-level Biology A specifications. It covers Module 3 Exchange and transport and Module 4 Biodiversity, evolution and disease. It is divided into two sections:

■ The **Content Guidance** provides key facts and key concepts, and links with other parts of the AS and A-level course. The links should help to show you how information in this module is useful preparation for other modules.

■ The **Questions & Answers** section contains two sets of questions, giving examples of the types of question to be set in the AS and A-level papers. There are some multiple-choice questions and some structured questions. The AS-style questions are followed by answers written by two students. These are accompanied with comments on the answers. The A-level-style questions are followed by model answers without comments.

This guide is not just a revision aid. It is a guide to the whole module and you can use it throughout the 2 years of your course if you decide to take the full A-level.

You will gain a much better understanding of the topics in Modules 3 and 4 if you read around the subject. I have suggested some online searches that you can do for extra information. Information you find online can help you especially with topics that are best understood by watching animations of processes taking place. As you read this guide remember to add information to your class notes.

The Content Guidance will help you to:

■ organise your notes and to check that you have highlighted the important points (key facts) — little 'chunks' of knowledge that you can remember

■ understand how these 'chunks' fit into the wider picture. This will help to support:
  – Module 2, which is covered in Student Guide 1 in this series
  – Modules 5 and 6, if you decide to take the full A-level course, which are covered in the third and fourth student guides in this series

■ check that you understand the links to the practical work, since you must expect questions on practical work in your examination papers. Module 1 lists the details of the practical skills you need to use in the papers.

■ understand and practise some of the maths skills that will be tested in the examination papers — look out for this icon for examples 🔢

The **Questions & Answers** section will help you to:

■ understand which examination papers you will take

■ check the way questions are asked in the AS and A-level papers

■ understand what is meant by terms like 'explain' and 'describe'

■ interpret the question material — especially any data that you are given

■ write concisely and answer the questions that the examiners set

Content Guidance

Module 3 Exchange and transport

■ Exchange surfaces

Key concepts you must understand

Humans are large, multicellular organisms. Although there are many organisms much larger than humans, there is a vast number that are smaller. Size is important when it comes to exchanging substances, especially oxygen and carbon dioxide, with the surroundings and then moving them around the body.

Figure 1 shows *Amoeba*, a single-celled organism with a body consisting of one mass of cytoplasm not subdivided into cells. *Amoeba* is non-photosynthetic and gains its energy by eating smaller organisms, such as bacteria. It lives in fresh water. Figure 1 also shows the exchange of gases that occurs between *Amoeba* and its surroundings. The cell surface membrane serves as the site of **gas exchange**, and its surface area is large enough — for the mass of the cytoplasm — to provide sufficient oxygen for respiration and for the removal of carbon dioxide. Movement of these gases occurs by diffusion.

Gas (gaseous) exchange The diffusion of oxygen and carbon dioxide between an organism and its environment.

Gas exchange surfaces Parts of the body where gas exchange occurs. For single-celled (unicellular) organisms the whole body is the gas exchange surface; fish have gills; mammals have lungs.

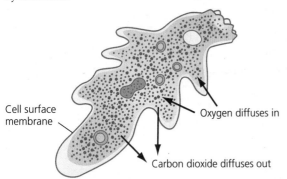

Cell surface membrane

Oxygen diffuses in

Carbon dioxide diffuses out

Figure 1 *Amoeba* is a single-celled organism that has a large surface area-to-volume ratio; it uses its body surface for gas exchange

Larger multicellular animals such as fish, insects and mammals, such as humans, do not have sufficient body surface to act as the site of gas exchange, because there is not enough surface to absorb the oxygen required. This is why animals have lungs or gills with large surface areas as sites of gas exchange. There are also specialised surfaces for exchange elsewhere in the body — for example, the lining of the gut has a large surface area for the absorption of digested food.

Animals have different demands for energy. Compare two fish — a monkfish that lives close to the bottom of the sea and moves sluggishly and a mackerel that swims very fast through open water. The mackerel has a much higher demand for oxygen and therefore has gills with a relatively higher surface area than the monkfish.

 It is difficult to calculate the surface area of animals and plants, but you need to know how the **surface area-to-volume ratio** changes as organisms increase in size. It helps to use cubes of different sizes to understand this principle. Table 1 shows what happens to the surface area-to-volume ratio as a cube increases in size.

Table 1

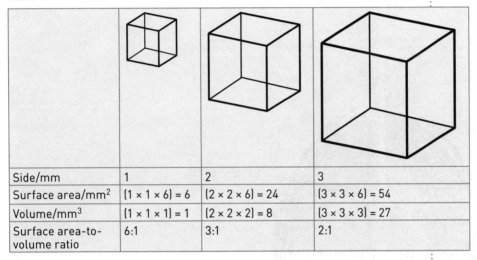

Side/mm	1	2	3
Surface area/mm²	(1 × 1 × 6) = 6	(2 × 2 × 6) = 24	(3 × 3 × 6) = 54
Volume/mm³	(1 × 1 × 1) = 1	(2 × 2 × 2) = 8	(3 × 3 × 3) = 27
Surface area-to-volume ratio	6:1	3:1	2:1

- Small organisms have a large surface area-to-volume ratio.
- Large organisms have a small surface area-to-volume ratio.
- Small organisms use their body surface for gas exchange, but larger organisms have specialised surfaces for exchange, for example gills and lungs.

Links

You can expect questions that involve calculations on both AS papers, so you may have to calculate surface area-to-volume ratios. To do this, divide the surface area by the volume — the resulting figure will represent how much surface area (in units of area, e.g. mm²) there is for every unit of volume (e.g. for every 1 mm³). Write this down as a ratio, for example $A{:}1$, where A is the number on your calculator. You may not get a whole number. If this is the case, round up or down to the lowest number of significant figures in the figures that you have used in the calculation. It is acceptable to write a surface area-to-volume ratio as something like 1.5:1.

Knowledge check 1

An organism has a cylindrical shape with the following dimensions: length = 21.00 mm, diameter = 5.30 mm. Calculate the SA:V ratio for this organism. Write down the formulae that you will use in your calculation and show your working.

Exchange in the mammalian lungs

Key concepts you must understand

Figure 2 shows the structure of the gas exchange system, consisting of the trachea and lungs. Figure 3 shows the tisses in more detail.

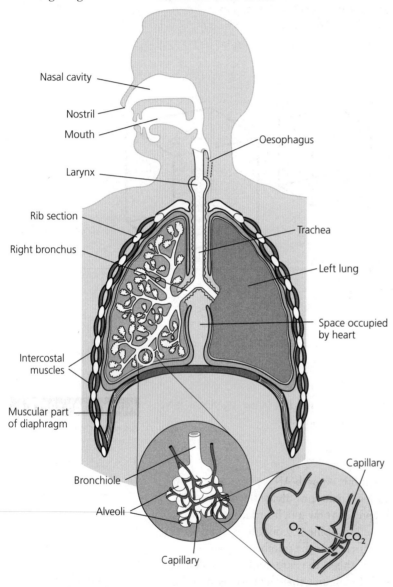

Figure 2 The gas exchange system

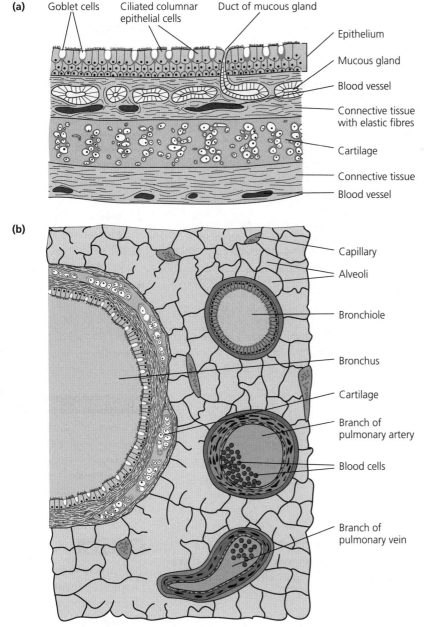

(a)
Goblet cells
Ciliated columnar epithelial cells
Duct of mucous gland
Epithelium
Mucous gland
Blood vessel
Connective tissue with elastic fibres
Cartilage
Connective tissue
Blood vessel

(b)
Capillary
Alveoli
Bronchiole
Bronchus
Cartilage
Branch of pulmonary artery
Blood cells
Branch of pulmonary vein

Figure 3 (a) Detail of the wall of the trachea; (b) distribution of tissues in the lungs

You should be clear about two different aspects of the gas exchange system:

■ **ventilation** — breathing air in and out of the lungs
■ gas exchange — diffusion of oxygen and carbon dioxide between air in the alveoli and the blood

Ventilation and gas exchange occur to provide cells with oxygen and remove the carbon dioxide that they produce. Oxygen is needed for respiration and carbon dioxide is its waste product. Do not confuse breathing and ventilation with respiration.

Ventilation Movement of air or water over a gas exchange surface.

Respiration Chemical process that occurs inside cells to transfer energy from molecules, such as glucose, triglycerides and amino acids, to ATP. Respiration may be aerobic or anaerobic.

Table 2 shows the distribution of the tissues and cells within the mammalian gas exchange system as shown in Figure 2.

Table 2 Components of the mammalian gas exchange system and their functions

Structure	Distribution in gas exchange system	Functions
Ciliated epithelium	Trachea, bronchi, bronchioles	Cilia move mucus up the airways
Goblet cells	Trachea, bronchi	Secrete mucus
Cartilage	Trachea, bronchi	Holds the airways open to allow easy flow of air
Smooth muscle (also known as involuntary muscle)	Trachea, bronchi and bronchioles	Contracts to narrow the airways
Elastic fibres	In all parts of the system, including alveoli	Stretch when breathing in; recoil when breathing out, helping to force air out of the lungs
Squamous epithelium	Alveoli	Thin, to give a short diffusion pathway for gas exchange; alveoli provide a large surface area
Capillaries	In all parts of the system — many around the alveoli	Provide a large surface for exchange between blood and air

Key facts you must know

Gas exchange in the alveolus

Alveoli are tiny air-filled sacs, adapted for the efficient exchange of gases by diffusion between the air and blood capillaries. There are two main ways in which alveoli are adapted for efficiency.

Short diffusion distance

Cells lining alveoli and blood capillaries are squamous epithelial cells, which are very thin. This allows easy diffusion of oxygen and carbon dioxide even though there are five cell membranes between the air and the haemoglobin inside red blood cells (Figure 4).

Steep concentration gradient

Breathing ventilates the alveoli, maintaining a high concentration of oxygen in alveolar air. Blood flows through capillaries in the lungs, bringing a constant supply of deoxygenated blood. Breathing and the flow of blood maintain a steep concentration gradient, so that diffusion of oxygen from the air into the blood is rapid. Exactly the opposite happens with carbon dioxide, although the concentration gradient is not as steep.

The lungs are ventilated by movements of the diaphragm and ribcage. You can take shallow breaths by movement of the diaphragm alone, but to breathe deeply you must use both the diaphragm and the ribcage.

Cartilage A type of connective tissue similar in structure to bone, but not as hard and much more flexible.

Exam tip

There are three types of muscle tissue in mammals:

- **skeletal** (also known as voluntary muscle) is attached to the skeleton by tendons
- **smooth** is found in internal organs such as the gut and gas exchange system
- **cardiac** is only found in the heart (p. 17)

Knowledge check 2

Write out the pathway taken by inspired air as it travels from the atmosphere to the site of gas exchange in the lungs.

Exam tip

Search online for photographs and electron micrographs of goblet cells so that you can recognise them. Notice the spelling of goblet — misspell it and you may not get the mark.

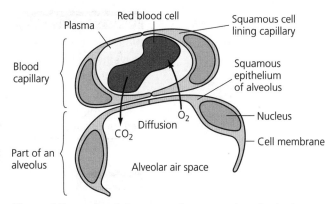

Figure 4 Features of the gas exchange surface in the lungs

Breathing in (inspiration/inhalation)

- Diaphragm muscles contract and pull the diaphragm down into the abdomen.
- External intercostal muscles contract and raise the ribcage.
- The volume of the thorax (chest cavity) *increases*.
- The pressure of air inside the lungs *decreases*.
- Air moves from the atmosphere to the lungs because the pressure in the atmosphere is greater than the air pressure in the lungs.

Breathing out (expiration/exhalation)

- Diaphragm muscles relax and the contents of the abdomen push the diaphragm upwards into the thorax.
- The ribs fall under gravity.
- The internal intercostal muscles may contract to lower the ribcage (e.g. in forced expiration).
- The volume of the thorax *decreases*.
- The pressure of air in the lungs is *greater* than the pressure of air in the atmosphere, so air is forced out.

Mammals ventilate their gas exchange surface using a bidirectional flow of air (in and out through the same openings — mouth and nose). This tends to reduce the loss of water vapour and heat from the body, which would happen if air exited the body via another route, such as a hole in the chest.

Lung volumes

- **Tidal volume** is the volume of air that you breathe in and then breathe out during one breath. At rest it is usually about $500\,cm^3$. It increases when you exercise, sing or play a wind instrument.
- **Vital capacity** is the volume of air you can force out after taking a deep breath. In young adult males it can be about $4.6\,dm^3$; in females, $3.1\,dm^3$. Athletes, singers and wind instrument players often have larger vital capacities.
- **Total lung volume** is the vital capacity plus the volume of air left in the lungs after you breathe out forcibly, which is about $1\,dm^3$. The total lung volume of young males can therefore be between $5\,dm^3$ and $6\,dm^3$.

Exam tip

Find some images of alveoli so that you can recognise them and their features. You should be able to identify the features of this gas exchange surface *that can be seen in images*. The most obvious of these are the short diffusion distance and the close proximity of capillaries.

Knowledge check 3

List three ways in which alveoli are adapted for gas exchange.

Knowledge check 4

a State how ventilation differs from gas exchange.
b Explain why respiration is not the same as breathing.

Exam tip

A good way to learn what happens when you breathe in and out is to put your hand on your ribcage and then later on your abdomen. Take some shallow breaths and then some deep ones.

Spirometer

Measurements of lung volumes such as tidal volume and vital capacity can be made with a spirometer. Figure 5 shows a typical spirometer.

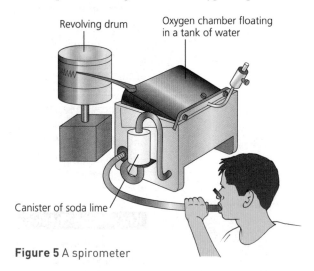

Figure 5 A spirometer

A pen fixed to the lid of the spirometer makes a trace on the paper fixed to the revolving drum. As the person breathes in, the lid falls and the pen makes a downward movement. As the person breathes out, the lid rises and the pen makes an upward movement. In Figure 6, the arrow indicates when the person starts breathing from the spirometer.

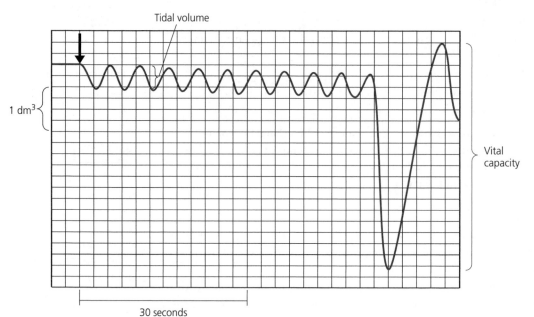

Figure 6 Measurements that can be made from a spirometer trace

Knowledge check 5

What happens to the movement of your ribcage and your abdomen when you take shallow and then deeper breaths?

Exam tip

Use a ruler on Figure 6 to verify these measurements. Similarly, when you have to analyse spirometer traces in the examination always use a ruler to make accurate data quotes.

You can see more spirometer traces in Question 6 of the AS paper (p. 71).

We can learn the following information from the spirometer trace shown in Figure 6:

- Tidal volume is the difference between peak and trough. Here, it is $0.5\,dm^3$ ($500\,cm^3$).
- Vital capacity is the total volume of air breathed out after taking a deep breath. Here it is $5\,dm^3$ ($5000\,cm^3$).
- Breathing rate is the number of peaks or troughs per minute. Here it is 6 in 30 seconds, which is 12 breaths min^{-1}.
- Oxygen consumption is the gradient of the trace. Here it is $125\,cm^3$ in 30 seconds ($250\,cm^3\,min^{-1}$).

Gas exchange in insects and fish

Key facts you must know

Insects have tiny holes called **spiracles** along both sides of their body, one pair in each segment (Figure 7). These open into tracheae that spread throughout the body from the thorax and abdomen and into the head (insects do not breathe through their mouths). The tracheae supply air sacs and branch into smaller and smaller tubes ending in tiny tracheoles, which are the site of gas exchange between air and tissues. The blood has no role in gas exchange.

At rest the tracheoles are partially filled with tissue fluid, but when an insect is active the fluid is reabsorbed and air can penetrate further down the tracheoles to increase the surface area. Spiracles can be opened and closed to control the flow of air; tracheae are held open by rings of chitin — similar in function to the cartilaginous rings of mammalian tracheae.

Insects ventilate their gas exchange system by contractions of the thorax and abdomen. If you watch a locust or grasshopper you can see these abdominal movements, which increase in frequency as the animal becomes more active or its body temperature increases. Insects often take in air through their thoracic spiracles and expel it through their abdominal spiracles, so achieving a unidirectional flow of air through the tracheal system.

Spiracle Small holes along the thorax and abdomen of an insect for inspiration and expiration during ventilation.

(a)

(b)

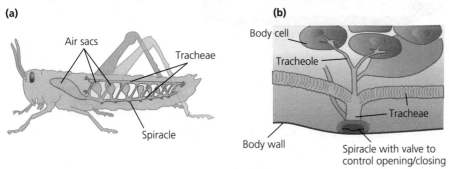

Figure 7 (a) The gas exchange system of a grasshopper. (b) On the outside of the body there are pairs of spiracles leading to tracheae, which take air into air sacs and tracheoles

The gills are the gas exchange system of bony fish, such as mackerel, trout and herring. Gill bars made of bone support gill filaments, which are covered in gill lamellae (also known as **secondary lamellae**), which are the site of gas exchange (Figure 8).

When a fish breathes it opens its mouth to take in water. Shortly afterwards a pair of openings appear just behind the head and water flows out of the body. These openings are each covered by an **operculum**, which is a bony plate covered in scales.

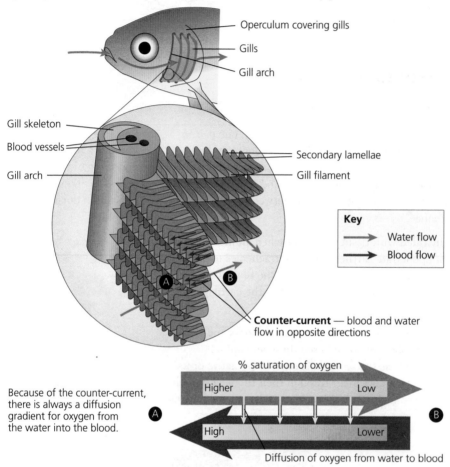

Because of the counter-current, there is always a diffusion gradient for oxygen from the water into the blood.

Figure 8 The gas exchange system of a bony fish

During inspiration the **buccal (mouth) cavity** expands to lower the pressure so water enters the mouth. When the buccal cavity is full, the mouth shuts and muscular contractions reduce its volume, increasing the pressure of water so it flows over the gills. The operculum then opens so that water can flow out. During inspiration, the pressure of water outside forces the operculum to close but the pressure inside the cavity decreases so even during this phase water flows across the gills to maintain a constant unidirectional flow of water.

Blood flows through the secondary lamellae in the opposite direction from the flow of water (Figure 8). This **countercurrent flow** makes sure that constant gradients for oxygen and carbon dioxide are maintained across the whole length of each secondary lamella.

Operculum The bony plate that covers the gills. It opens to allow water to move out of the gill cavity.

Exam tip

The secondary lamellae form the gas exchange surface. The gills are the organs in which gas exchange occurs.

Knowledge check 6

State the gas exchange surfaces of insects, bony fish and mammals.

Buccal cavity The space between the mouth and the start of the intestine.

Countercurrent flow Movement of two liquids in opposite directions to maximise exchange between them.

Exam tip

Find photographs of the gas exchange systems of insects and fish to help you understand how they function. List the similarities and differences between them using a table.

Animals have other types of exchange surface: the intestine for the absorption of nutrients and excretory organs, such as kidneys, that reabsorb useful substances so they are not lost from the body. Although plants rarely have specialised structures for gas exchange (see Exam tip on p. 16), they do have root hairs for absorbing water and ions. Exchange surfaces have the following common features:

- large surface area to absorb as much as possible
- thin surface to give a short diffusion distance
- methods to maintain steep concentration gradients, for example by transporting absorbed substances away from the exchange surface

Summary

- The ratio of surface area to volume decreases as cells and whole organisms increase in size. As a cell increases in size its volume increases as the cube of its linear dimensions, but the surface area only increases as the square. This explains why almost all cells and single-celled organisms are small and why multicellular organisms have specialised surfaces for gas exchange.
- The features of an efficient gas exchange surface are: large surface area; short distance separating the blood from the environment (water or air); good blood supply (but not in insects); and a method of moving water or air over the exchange surface to maintain constant concentration gradients.
- Alveoli are lined by a thin squamous epithelium; have elastic fibres to allow expansion and elastic recoil; are surrounded by many capillaries; and are ventilated by movements of the diaphragm and ribcage.
- Rings of cartilage in the trachea keep it open so there is little resistance to air flow. Elastic fibres in the airways allow them to stretch during inspiration and recoil during expiration to force air out. Smooth muscle in the walls of the airways contract to maintain tension and reduce the diameter.
- The airways are lined by a ciliated epithelium; goblet cells secrete mucus to trap bacteria, dust and other fine particles; cilia move the mucus towards the throat.

- During inspiration, the diaphragm and external intercostal muscles contract to increase the volume of the thorax and reduce air pressure in the lungs. Air moves in as atmospheric pressure is greater than air pressure in the lungs. During expiration, the diaphragm relaxes and the rib cage falls. Internal intercostal muscles may contract. The volume of the thorax decreases and air pressure in lungs increases, so is greater than atmospheric pressure. Air is breathed out.
- Tidal volume (TV) is the volume of air breathed in with each breath. Vital capacity (VC) is the maximum volume of air that can be breathed out. TV changes with behaviour, for example it increases during exercise and singing; VC changes with age and training.
- A spirometer is used to measure TV, VC and breathing rate (as breaths min^{-1}). Rates of oxygen uptake in $cm^3\,min^{-1}$ or $dm^3\,min^{-1}$ are measured by the gradual decrease in total volume of the spirometer.
- Insects breathe by contracting and expanding their thorax and abdomen to draw air in and out of spiracles. The air flows through tracheae to supply tracheoles, where gas exchange occurs between air and tissues — the blood has no role.
- Secondary lamellae form the gas exchange surface of bony fish. These are ventilated by movement of water over the gills in an opposite direction to the flow of blood. This countercurrent flow maintains a constant difference in concentrations across the length of the secondary lamellae.

■ Transport in animals

Key concepts you must understand

Substances such as oxygen, carbon dioxide and absorbed food have to be moved around the body. In single-celled organisms, such as *Amoeba*, substances can pass by diffusion or be carried in cytoplasm as it flows within the organism. Diffusion does not work in a large organism because distances are too great, and oxygen cannot be supplied fast enough from the lungs to cells elsewhere in the body; a transport system is needed.

In this module you study three systems — blood, xylem and phloem. All three are examples of **mass flow** — the movement of a fluid through a system of tubes in one direction. Table 3 compares the transport mechanisms in mammals and terrestrial flowering plants.

> **Exam tip**
> Notice that plants do not have a transport system for gases. Diffusion through the extensive air spaces inside leaves is sufficient. Other areas such as the centres of the trunks and roots of trees and shrubs do not have many living cells and they have very low metabolic rates.

Table 3 Comparing transport systems in mammals and terrestrial flowering plants

Feature	Mammals	Terrestrial flowering plants
Transport system	Circulatory system: heart + blood vessels + blood	Xylem and phloem
Gas exchange surface	Alveoli in the lungs	All cell surfaces that are in contact with the air, for example palisade and spongy mesophyll cells in leaves
Transport of oxygen	Oxygen in combination with haemoglobin	Oxygen and carbon dioxide diffuse through air spaces between cells
Transport of carbon dioxide	Carbon dioxide in blood plasma as HCO_3^- and in combination with haemoglobin	
Transport of carbohydrate	Glucose in solution in blood plasma	Sucrose in solution in phloem sap
Transport of water	Most of the blood plasma is water	In the xylem sap
Force to move fluids	Heart	Xylem — transpiration pull Phloem — active pumping of sugars into the phloem and hydrostatic pressure

Animals are multicellular — their bodies are made of many cells. Some small animals rely on diffusion alone to transfer oxygen, carbon dioxide, small molecules absorbed after digestion and waste substances around the body.

Types of circulatory system in animals
Key facts you must know

Insects have an **open circulatory system** in which blood is pumped by a heart into spaces that extend throughout the body and bathes the tissues directly; there are no blood vessels. Other animals, such as annelid worms (e.g. earthworms) and vertebrates, have a **closed circulatory system** in which blood flows through the body enclosed within blood vessels. Substances are exchanged between blood and tissues across the walls of capillaries — the tiniest blood vessels.

Closed circulations have three important components:

■ the blood — red blood cells, white blood cells, platelets and plasma
■ blood vessels — arteries, arterioles, capillaries, venules and veins
■ the heart — the pump for circulating blood through the vessels

> **Knowledge check 7**
> Distinguish between open and closed circulatory systems.

Figure 9 shows simple views of fish and mammalian circulatory systems. Fish have a single circulation — the blood passes from the heart to the gills and then immediately to the rest of the body. Blood passes through the heart once during a complete circulation of the body. Mammals have a double circulation in which the blood passes through the heart twice in one complete circulation of the body. The circuit from the heart to the lungs and back is the **pulmonary circulation**. The circuit from the heart to the rest of the body and back is the **systemic circulation**. Find these two circulations in Figure 9(b).

Knowledge check 8

Distinguish between single and double circulatory systems.

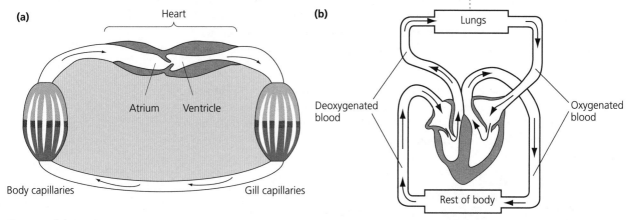

Figure 9 (a) The fish circulatory system. Follow the pathway from the gills and back to see why it is called a single circulation. (b) The mammalian circulatory system. Follow the pathway from the lungs and back to see why it is called a double circulation

The mammalian heart
Key facts you must know

The heart is a muscular pump. It is made of cardiac muscle and is described as **myogenic** (it stimulates itself to beat). You should be familiar with the following features of the heart:

- There are four chambers — two atria and two ventricles.
- There are two pumps working in series — the right side of the heart pumps deoxygenated blood to the lungs in the pulmonary circulation; the left side pumps oxygenated blood to the rest of the body through the systemic circulation.
- The left and right atria have thin walls because they pump blood into the ventricles, which are only a short distance from the atria.
- The left and right ventricles have thick walls because they pump blood a greater distance and against a greater resistance than the atria.
- There are valves in the heart to prevent backflow and to ensure the blood follows the correct pathway.
- The volume of blood ejected by each chamber is the same during one beat, but the volume can change from beat to beat in response to the body's demand for oxygen.

Myogenic Muscle contraction initiated by the muscle itself, not by an external influence such as nerve impulses.

Knowledge check 9

Distinguish between systemic and pulmonary circulations.

Stroke volume is the volume of blood ejected from each ventricle during one beat. At rest it can be about 70 cm³. **Cardiac output** is the volume of blood ejected from each of the ventricles during 1 minute:

cardiac output = stroke volume × heart rate

At rest it can be about 70 × 72 = 5040 cm³ = 5.04 dm³ per minute.

Drawings and diagrams of the heart usually show it viewed from the front of the body. Figure 10 shows the external structure of the heart with the major blood vessels. Note that **coronary arteries** supply the heart muscle with oxygenated blood. They branch from the base of the aorta. Figure 11 shows the internal structure of the heart.

Table 4 shows the functions of the chambers of the adult human heart. The ventricles are thicker than the atria because they generate greater pressures. The left ventricle is thicker than the right ventricle because a high pressure is needed to overcome the resistance of the systemic circulation, which is much higher than the resistance in the pulmonary circulation.

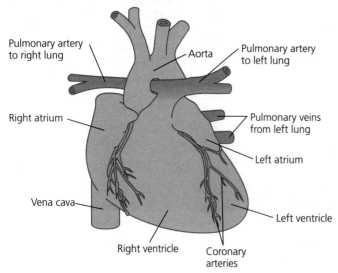

Figure 10 External view of the heart

The lungs are a spongy tissue. Blood fills most of the capillaries in the lungs. A low blood pressure in the pulmonary circulation ensures that fluid does not leak out of the pulmonary capillaries, causing fluid to accumulate in the alveoli.

Control of the heart

The heartbeat is controlled by the sino-atrial node (SAN) in the right atrium. The SAN is a special region of muscle cells that emits electrical pulses similar to those that pass along nerve cells. These travel across the muscle in the atria and cause them to contract together. The electrical impulses are prevented from reaching the ventricle muscles directly by a ring of fibrous tissue between the atria and the ventricles.

> **Knowledge check 10**
>
> The cardiac output during exercise is 7.50 dm³ min⁻¹ and the heart rate is 90 beats min⁻¹. Calculate the stroke volume. Write down the formula that you use and show your working.

> **Knowledge check 11**
>
> Describe the pathway that blood takes between the vena cava and the aorta.

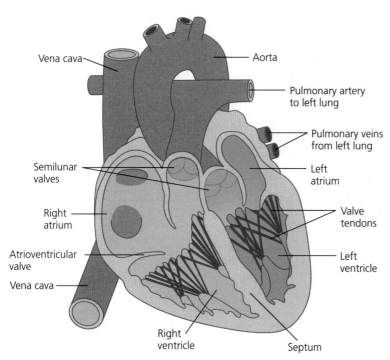

Figure 11 Vertical section through the heart

Table 4 The functions of the four chambers of the mammalian heart

Chamber of the heart	Receives blood from...	Pumps blood to...
Right atrium	...the body through vena cava	...the right ventricle
Right ventricle	...the right atrium	...the lungs through the pulmonary artery
Left atrium	...the lungs through pulmonary veins	...the left ventricle
Left ventricle	...the left atrium	...the body though the aorta

The atrioventricular node (AVN) is in the central septum at the junction between the atria and ventricles. The AVN delays the impulses so that they reach the ventricles after these chambers have filled with blood from the atria. Impulses are relayed by the AVN along Purkyne tissue, which conducts to muscles at the base of the ventricles so that this area contracts first. This forces blood from the bottom of the ventricles upwards into the arteries. The rest of the ventricle muscles contract so that the ventricles empty completely.

The cardiac cycle

The **cardiac cycle** describes the sequence of changes that occurs in the heart during one heartbeat. The spread of impulses from the SAN starts a series of changes that make up the cardiac cycle. Imagine that the individual drawings in Figure 12 are still pictures from a film of the heart beating. The small arrows in the heart show where the blood is flowing at each stage. Figure 13 shows the changes in blood pressure in the left atrium, left ventricle and aorta.

Exam tip

These electrical pulses are also described as waves of depolarisation. Nerves that supply the heart alter the rate of contraction; they do not stimulate it to contract.

Knowledge check 12

List the functions of the SAN, AVN and Purkyne tissue.

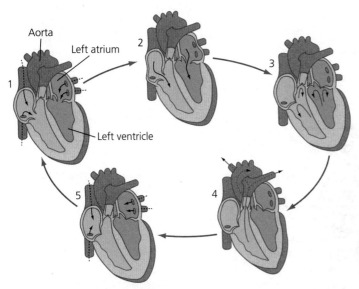

Figure 12 The cardiac cycle

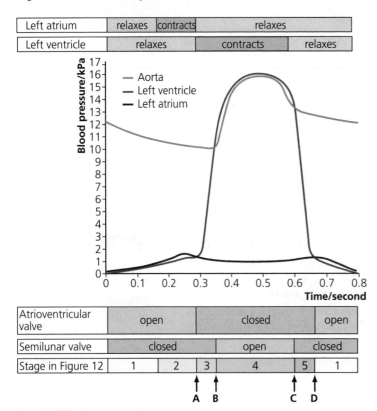

Left atrium	relaxes	contracts	relaxes	

Left ventricle	relaxes	contracts	relaxes

Atrioventricular valve	open	closed	open

Semilunar valve	closed	open	closed

Stage in Figure 12	1	2	3	4	5	1

A   B   C   D

Figure 13 Contraction occurs in stages 2, 3 and 4 — this is systole. Relaxation occurs in stages 5 and 1 — this is diastole

It takes 0.8 seconds to complete this cardiac cycle. This means that there are 60/0.8 = 75 beats per minute.

Ask your teacher for three sheets of transparent acetate paper. Draw out the axes of Figure 13 on each piece of acetate. It is a good idea to make a large copy. Copy the curve for the left atrium onto one sheet; the curve for left ventricle on the second sheet and the curve for the aorta on the third. Read the following bullet points carefully and follow the changes on your acetate sheets. Then overlay them and find the 'crossing points' when valves open and close.

Notice the following:

■ The blood pressure in the atria and the ventricles falls near to 0 kPa during each cycle because there are times when there is little blood in these chambers.

■ The pressure in the aorta does not fall below about 10 kPa. Its wall stretches as blood surges into it from the left ventricle, and then recoils to maintain the blood pressure and to keep blood flowing.

■ When the pressure in the left ventricle is greater than that in the left atrium, the atrioventricular valve closes to stop blood flowing back into the atrium. The tendons at the base of the valve stop the blood 'blowing back' into the atrium. This occurs at point A in Figure 13.

■ When the pressure in the ventricle is greater than that in the aorta, the semilunar valve opens and the blood flows from the ventricle to the aorta. This occurs at point B in Figure 13.

■ When the pressure in the aorta is greater than that in the ventricle, the semilunar valve fills with blood, closes off the aorta and prevents backflow. This occurs at point C in Figure 13.

■ When the pressure in the left atrium is greater than the pressure in the left ventricle, the atrioventricular valve opens and blood flows from the left atrium into the ventricle. This occurs at point D in Figure 13.

Now position a ruler on Figure 13 over the vertical axis at time 0.

■ Move the ruler across the graph to the right and follow the changes to the blood pressure in the ventricle. Look at the stages (1–5) in Figure 12 to follow the changes to the left ventricle. Move the ruler back to 0 and repeat the procedure, this time following the changes in the left atrium, and then again for the changes in the aorta.

■ Think about the opening and closing of the semilunar and atrioventricular valves. Look at the blood pressure either side of points A, B, C and D in Figure 13 and at Figure 12 to check whether these valves are opening or closing at each point in the cycle.

Electrocardiograms

One way to check on the health of the heart is to record its electrical activity by using electrocardiography. Electrodes are attached to the body to give an electrocardiogram (ECG). Figure 14 shows a normal ECG trace. Abnormal ECGs are described in Table 5 and there are some abnormal ECG traces in Question 7 on p. 75.

Exam tip

For further information, look at animations of the cardiac cycle that show the changes that occur in the heart as well as the pressure changes shown in Figure 13.

Knowledge check 13

Draw a graph like that in Figure 13, but show the changes in the right side of the heart and in the pulmonary artery.

Knowledge check 14

Suggest the advantages of having a double circulation rather than a single circulation.

Exam tip

You can expect to be asked about the cardiac cycle graph, so it is well worth studying Figures 12 and 13 in some detail.

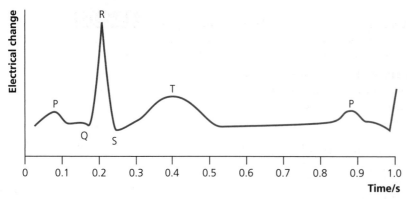

Figure 14 A normal ECG trace

Table 5 Abnormal ECGs

Condition	Differences with normal ECG
Tachycardia	Fast heart rate; the distances between the P waves are shorter than in Figure 14
Bradycardia	Slow heart rate; the distances between the P waves are longer than in Figure 14
Atrial fibrillation	Fast and irregular heart rate; small waves in between PQRS waves
Ventricular fibrillation	No distinct PQRS or T waves; high rate of waves of varying amplitude
Ectopic heartbeat	Occasional heartbeats have a different pattern of PQRS waves; each such heartbeat is followed by a long relaxation phase

Links

If you dissect a heart as part of your practical work, do not expect the interior to look like the diagrams. Use diagrams to anticipate what to expect, but draw what you see. You may be asked to measure the thickness of the walls of the four chambers and record them on your drawing, and explain the differences in thickness.

Interpreting ECGs requires detailed knowledge far beyond this level — but you should find examples of ECGs showing the five conditions listed in Table 5. The heart rate is influenced by nerves and by hormones, details of which are in Module 5.

Blood vessels

Key concepts you must understand

An important function of the blood is to transport gases (oxygen and carbon dioxide), nutrients, hormones and waste products (e.g. urea). During each circuit of the body, blood flows through capillaries in the lungs and other organs, such as the stomach, liver and kidneys. Figure 15 shows the path taken by blood as it flows through a capillary bed in an organ. Capillaries are the **exchange vessels** of the circulatory system, where substances pass into and out of the blood.

Blood is supplied to an organ by an artery. It then flows through arterioles and then capillaries where exchanges occur between blood and tissue fluid. Blood drains through venules and then veins to return to the heart. Capillaries are very small, which is why the central area in Figure 15 is shown magnified.

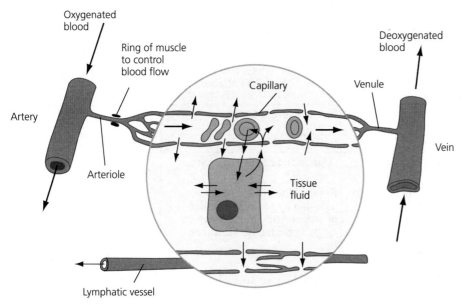

Figure 15 Blood flow through a capillary bed

Exam tip

Look at Figure 17 (p. 24) to see how blood pressure changes as it flows through the vessels in Figure 15.

Key facts you must know

Figure 16 shows cross-sections of an artery and a vein. Table 6 compares the structures and functions of arteries and veins.

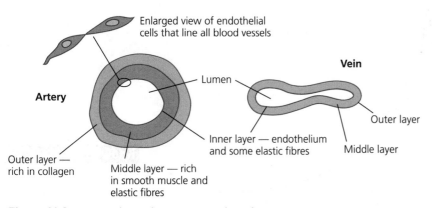

Figure 16 Cross-sections of an artery and a vein

Exam tip

You need to recognise arteries and veins in photographs and microscope slides and be able to draw them under the microscope.

As blood flows around the body, its pressure changes. Blood pressure is necessary — without it the blood would not flow through the blood vessels. The vessels present a resistance to blood flow and the contraction of the heart raises the pressure of the blood, forcing it through the circulation. There is a high pressure in the arteries so that blood is delivered efficiently to organs. Blood then flows through many smaller blood vessels — arterioles — before flowing through an even larger number of capillaries. As the diameter of the vessels decreases, the resistance to flow increases significantly.

Arterioles are surrounded by smooth muscle, which can contract to reduce the size of the lumen, so that less blood flows to capillaries. This allows blood to be diverted elsewhere, for example from skin and gut to muscles during exercise. There

Table 6 The structure and functions of arteries and veins

Feature	Artery	Vein
Width of wall	Thick	Thin
Components of wall	Lined by endothelium; rest of wall contains smooth muscle, elastic fibres, collagen	As vein, but less smooth muscle, elastic fibres and collagen
Semilunar valves	No	Yes, to prevent backflow
Blood pressure	High	Low
Direction of blood flow	Heart to organs and tissues	Organs and tissues to heart
Function	Elastic fibres recoil to maintain high pressure in arteries and overcome resistance of the circulation system	Return blood to the heart; assisted by squeezing action of surrounding muscle, which helps to push blood towards heart

Exam tip

There is very little smooth muscle in the walls of veins — certainly not enough to force blood back to the heart. The contraction of muscles around veins is *not* by peristalsis, which refers to the waves of muscle contraction that occur along the oesophagus to move food to the stomach.

is only enough blood to fill 25% of the capillaries at any one time, so arterioles play an important function in controlling the flow of blood to tissues, rather like taps controlling the flow of water. Arteries and veins may be metres in length (think of a blue whale or a giraffe); arterioles, capillaries and venules are only a few millimetres in length. Changes in blood pressure in the systemic circulation are shown in Figure 17.

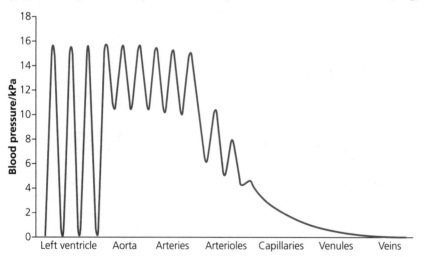

Figure 17 The changes in blood pressure at different places in the systemic circulation (*x*-axis not drawn to scale)

Knowledge check 15

List the names of the main arteries and veins in the systemic circulation.

The graph in Figure 17 shows the pressure changes in three successive heart beats and the effect of these in the arteries and other vessels. Notice that there are three peaks and troughs in the left ventricle, which are repeated in the aorta and the arteries. The rise and fall in blood pressure is greatest in the left ventricle. This is then reduced in the aorta and main arteries and becomes much smaller in the arterioles and capillaries, where the blood pressure decreases considerably. Blood pressure is lowest in the venules and veins.

From Figure 17 you can see why arteries have thick, muscular and elastic walls (to withstand high blood pressure), arterioles have muscular walls (to damp down blood pressure) and veins have thin walls (as the blood has a low pressure). Capillaries have thin walls because they are exchange vessels (see Figures 15 and 18). Figure 19 shows a cross-section of a capillary.

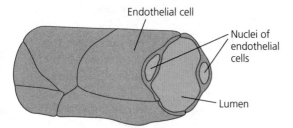

Figure 18 A capillary lined by endothelial cells

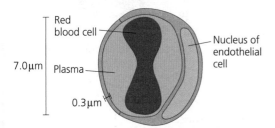

Figure 19 Cross-section of a capillary. Notice that red blood cells just fit inside the vessel and that the endothelial cells are thin, to help the exchange of substances with the blood

Exchanges that occur as blood flows through capillaries include:
- oxygen diffusing out of the blood and into tissue fluid
- carbon dioxide diffusing from tissue fluid into the blood
- water and dissolved substances, such as glucose and amino acids, being forced out by the pressure of the blood

Water and substances that are forced out by high pressure are **filtered** from the blood. As blood flows through the capillaries its **hydrostatic pressure** decreases, which makes it possible for water to drain back into the blood plasma by osmosis. This is because the blood contains solutes, such as albumen, which give the blood plasma a lower water potential than the tissue fluid. This lower water potential is also known as the blood **oncotic pressure**. Albumen is a large protein molecule that rarely leaves the blood. It is made by the liver to maintain the blood's oncotic pressure, so helping to maintain the water content of the blood so that it does not become too viscous and flow less easily.

Within tissues are small, blind-ended tubes called lymphatic vessels, which act as a 'drainage system'. Tissue fluid flows into the small lymphatic vessels and then flows slowly towards larger lymphatic vessels that empty into the blood near the heart. The fluid inside lymphatic vessels is lymph, and is similar in composition to tissue fluid.

At intervals along the larger lymphatic vessels are lymph nodes. These contain lymphocytes, some of which flow into the blood via the lymph. Lymph also drains fat from the small intestine so that after a meal it often appears as a white suspension.

Table 7 The composition and functions of three body fluids

Feature	Blood	Tissue fluid	Lymph
Where found	In blood vessels	Surrounding cells	In lymphatic vessels
Components:			
red blood cells	✓	✗	✗
white blood cells	✓	✓ (some)	✓ (some)
fats	✓ (as lipoproteins)	✓	✓ (especially after a meal)
glucose	✓	✓	Very little
proteins	✓	✓ (some)	✓ (some, e.g. antibodies)
Functions	Transport	Bathes cells — all exchanges between blood and cells occur through tissue fluid	Drains excess tissue fluid, preventing a build-up leading to oedema

Transport of gases

Key concepts you must understand

Cells respire. To respire aerobically they need a supply of oxygen and a way for waste carbon dioxide to be removed. If we only had a watery fluid to transport oxygen and carbon dioxide, we would not be able to carry these gases efficiently. Oxygen is not very soluble in water. The volume of oxygen that will dissolve in water is about $0.3\,cm^3$ per $100\,cm^3$. Blood can carry $20\,cm^3$ of oxygen per $100\,cm^3$. Carbon dioxide is much more soluble in water than oxygen is — about $2.6\,cm^3$ per $100\,cm^3$ can be transported in solution. However, blood can carry 50–$60\,cm^3$ of carbon dioxide per $100\,cm^3$. How is this possible?

Haemoglobin (known as Hb for short) transports almost all the oxygen in the blood and some of the carbon dioxide. Each red blood cell has about 280 million molecules of haemoglobin. You should revise the structure of haemoglobin, which you studied in Module 2 (Student Guide 1).

Haemoglobin is a protein composed of four sub-units, each one containing a haem group, which binds oxygen. Each haem binds one molecule of oxygen (O_2), so each haemoglobin molecule can bind to four molecules of oxygen ($4O_2$) to form oxyhaemoglobin:

$$\begin{array}{ccccc} \text{Hb} & + & 4O_2 & \rightleftharpoons & \text{HbO}_8 \\ \text{haemoglobin} & + & \text{four molecules of oxygen} & \rightleftharpoons & \text{oxyhaemoglobin} \end{array}$$

When carbon dioxide dissolves in water, much of it reacts to form carbonic acid, which dissociates (breaks up) to form hydrogen ions and hydrogencarbonate ions:

$$\begin{array}{ccccccccc} CO_2 & + & H_2O & \rightleftharpoons & H_2CO_3 & \rightleftharpoons & H^+ & + & HCO_3^- \\ \text{carbon} & + & \text{water} & \rightleftharpoons & \text{carbonic} & \rightleftharpoons & \text{hydrogen} & + & \text{hydrogencarbonate} \\ \text{dioxide} & & & & \text{acid} & & \text{ions} & & \text{ions} \end{array}$$

Blood can carry much more carbon dioxide than water can because there is a fast-acting enzyme inside red blood cells. This enzyme, **carbonic anhydrase**, catalyses

Exam tip

Draw a large diagram of a capillary surrounded by some cells and indicate all the exchanges that occur between blood and cells via tissue fluid. This might help you to understand this topic.

Exam tip

Haemoglobin is a superb example of many principles covered in Modules 2, 3 and 4. Read about it to learn much about protein structure, transport, genetics, inherited diseases and many other aspects of biology.

Exam tip

Haem is an iron-containing substance. Iron gives haem its oxygen-binding property. It is the prosthetic group of each of the four polypeptides that make up a haemoglobin molecule.

the formation of carbonic acid, which immediately dissociates to form hydrogen ions and hydrogencarbonate ions. Large quantities of hydrogencarbonate ions are carried in the blood plasma in association with sodium ions. Some carbon dioxide also attaches to the $-NH_2$ groups (amino groups) at the ends of the polypeptides in haemoglobin to form carbaminohaemoglobin, and some remains in solution (as carbon dioxide) in the plasma.

Key facts you must know

In the lungs, deoxygenated blood flows very close to alveolar air. The air in the alveoli is rich in oxygen. This richness is expressed as its partial pressure (abbreviated to pO_2), which is the part of the air pressure that oxygen exerts. Approximately 13–14% of the air inside the alveoli is oxygen. The total air pressure is about 100 kPa, so the partial pressure of oxygen in the alveoli is close to 13–14 kPa.

Deoxygenated blood flowing into the lungs has a low concentration of oxygen. Alveolar air is rich in oxygen, so a concentration gradient exists between the air and the blood, and oxygen diffuses from the air in the alveoli into the blood.

The oxygenated blood leaving the lungs carries almost its full capacity. Tissues such as muscle tissues and those in the gut, liver and kidney use oxygen in respiration. The concentration of oxygen in these tissues is low, equivalent to a partial pressure of approximately 5.0 kPa. This means that oxygen diffuses from the blood into the tissues down a concentration gradient. Blood loses about 30% of the oxygen it carries as it flows through the tissues when you are at rest (not doing any exercise).

To find out how much oxygen is transported by haemoglobin, small samples of blood are exposed to gas mixtures with different concentrations of oxygen. The volume of oxygen absorbed by the haemoglobin in each sample of blood is determined and expressed as a percentage of the maximum volume that haemoglobin absorbs. The results are shown as an oxygen–haemoglobin dissociation curve (Figure 20). The concentration of oxygen is shown as the partial pressure of oxygen in the gas mixture.

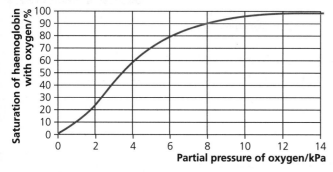

Figure 20 The results of an investigation into the effects of different air mixtures on the saturation of haemoglobin with oxygen

> **Exam tip**
>
> You can read more about the transport of carbon dioxide in the next section and see the changes that occur in Figures 21 and 22 (p. 29). Do not confuse carbaminohaemoglobin with carboxyhaemoglobin, which forms when *carbon monoxide* combines permanently with haemoglobin.

> **Exam tip**
>
> The partial pressure of oxygen (pO_2) is the pressure exerted by oxygen in a mixture of gases. Oxygen forms 21% of the atmosphere. The pO_2 of the atmosphere is approximately a fifth of one atmosphere (1 atm) or a fifth of 101.3 kPa using the SI unit of pressure. The pO_2 is 21.3 kPa. The pO_2 in alveolar air is much less because the concentration of oxygen in alveolar air is less as it is constantly being absorbed by the blood.

The line of best fit drawn on the graph is known as the oxygen–haemoglobin dissociation curve. The sigmoid shape (or S shape) of the line is important. Remember that the graph shows the results of an experiment on blood carried out in the laboratory. How does this relate to the actual situation in the mammalian body? If we take some figures from the graph, we should be able to see how to use this information. Table 8 shows the saturation of haemoglobin with oxygen at different partial pressures. The partial pressures chosen correspond with those in different parts of the circulation.

Table 8

Site in the body	Partial pressure of oxygen/kPa	Saturation of haemoglobin with oxygen/%
Lungs	13.0	98
Tissues, including muscles at rest	5.0	70
Muscles during strenuous exercise	3.0	43
Muscles at exhaustion	2.0	20
Placenta	4.0	60

We can see from Figure 20 and Table 8 that haemoglobin is:

- nearly fully saturated at the partial pressure of oxygen in the lungs
- about 70% saturated at the partial pressure in the tissues
- about 45% saturated in areas with very low partial pressures of oxygen, such as actively respiring muscle

As blood flows through capillaries, oxyhaemoglobin releases oxygen *in response to the low concentration of oxygen in the tissues*. Haemoglobin has a high affinity for oxygen at high partial pressures and a low affinity at low partial pressures.

Gas exchange between maternal blood and fetal blood occurs across the placenta. In order to be nearly fully saturated with oxygen, fetal haemoglobin must have a higher affinity for oxygen than adult haemoglobin. When experimenters investigated fetal blood using the same method as used to give the results shown in Figure 20, they found that the dissociation curve was *to the left* of the curve for adult haemoglobin.

The partial pressure of oxygen at the placenta is about 4.0 kPa. At this partial pressure, adult blood is close to 60% saturated, which means that oxyhaemoglobin gives up its oxygen to the surrounding tissues in the placenta. Fetal haemoglobin is about 80% saturated at this partial pressure, so it absorbs much of the oxygen that is released by the mother's blood. Tissues in the placenta absorb the rest.

Transport of carbon dioxide

Key facts you must know

Figure 21 shows the events that occur as carbon dioxide diffuses into the blood in tissues.

When the changes shown in Figure 21 occur, hydrogen ions (H^+) are released. These would reduce the pH of blood cells if left unchecked. Haemoglobin binds these hydrogen ions to become haemoglobinic acid (HHb). In this way haemoglobin acts as a **buffer** to the change in pH, stopping it from decreasing.

> **Knowledge check 16**
>
> Copy Figure 20 and draw on a dissociation curve for fetal haemoglobin.

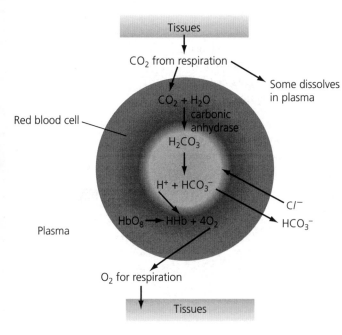

Figure 21 Most of the carbon dioxide diffuses into red blood cells and is converted to hydrogencarbonate ions by the action of carbonic anhydrase. Some carbon dioxide dissolves in the plasma. Hydrogencarbonate ions are exchanged with chloride ions from the plasma through specific channel proteins

The concentration of hydrogencarbonate ions in red blood cells is higher than in the plasma. These ions diffuse through channel proteins in the cell membranes out of the red blood cells into the plasma. Cell membranes are impermeable to hydrogen ions so they cannot diffuse out as well and besides, as shown in Figure 21, they combine with haemoglobin. Instead, the loss of negatively charged ions (anions) is counteracted by the inflow of chloride ions through the same channel proteins to maintain electrochemical neutrality. This exchange of chloride ions for bicarbonate ions is the **chloride shift**.

When the blood reaches the lungs, these changes go into reverse, as shown in Figure 22.

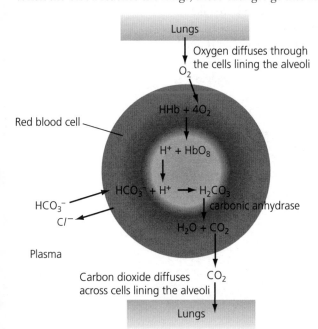

Figure 22 These events occur as blood flows through capillaries in the alveoli, so that carbon dioxide diffuses into alveolar air. The chloride shift goes into reverse

The Bohr effect

When carbon dioxide diffuses into the blood, it stimulates haemoglobin to release even more oxygen than it would if it only responded to the low concentration of oxygen. This can be explained by the fact that when haemoglobin accepts hydrogen ions to become haemoglobinic acid (HHb) it stimulates the molecule to give up oxygen (Figure 23).

When the experiment with gas mixtures was repeated using carbon dioxide as well, it was discovered that carbon dioxide *decreased* the affinity of haemoglobin for oxygen. Figure 23 shows the effect when the results are plotted on a graph.

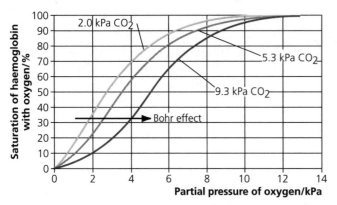

Figure 23 The Bohr effect. When carbon dioxide was added to the gas mixtures, haemoglobin became less saturated with oxygen

Exam tip

Always use a ruler to help you understand what is shown by graphs and to take accurate data quotes to illustrate your exam answers.

This shift of the curve to the right is known as the Bohr effect, after the Danish scientist who discovered it. The best way to understand the Bohr effect is to take some figures from the graph. Table 9 shows the effect of increasing the partial pressure of carbon dioxide on the saturation of haemoglobin. The figures are taken for the *same* partial pressure of oxygen ($pO_2 = 3.0\,kPa$, which corresponds with that in the tissues when they are respiring actively). Notice that the haemoglobin is much less saturated with oxygen when the pCO_2 is $9.3\,kPa$. As blood flows through the capillaries it gives up even more oxygen than it would have done if there was less carbon dioxide present.

Table 9

Partial pressure of carbon dioxide/kPa	Percentage saturation of haemoglobin with oxygen at pO_2 of 3.0
2.0	55
5.3	40
9.3	20

Links

If fetal haemoglobin has a higher affinity for oxygen than adult haemoglobin, why don't we keep it throughout life? Unfortunately, because it has a higher affinity, it does not release its oxygen so well in respiring tissues. After birth, tissues need a much greater supply of oxygen and this is provided by adult haemoglobin — a switch occurs during the first few years of life. Sometimes this switch does not happen, with the result that the tissues are starved of oxygen. This is what happens in people who have thalassaemia, a genetic disease that occurs most frequently in people from countries around the Mediterranean. Thalassaemia originated as a mutation in one of the genes that codes for haemoglobin. You will learn more about mutations in Module 6.

Summary

- Multicellular animals need transport systems because they are large, often active with a high metabolic rate and have a low SA:V ratio.
- Insects have an open circulatory system with a heart that pumps blood into a body space, not into blood vessels. Fish and mammals have a closed circulatory system with blood enclosed in vessels. Fish have a single circulatory system with blood travelling once through the heart in one complete circulation; mammals have a double circulatory system with one circuit to the lungs and another to the rest of the body.
- The mammalian heart has four chambers; atria are thin walled as they pump blood a short distance to thicker-walled ventricles. The left ventricle has a thicker wall as it pumps blood against a greater resistance than the right, which pumps blood to the lungs.
- The cardiac cycle is the sequence of contractions and relaxations of the four chambers; changes in blood pressure in the left atrium, left ventricle and aorta are shown on a graph. The semilunar valves open during ventricular contraction; the atrioventricular valves open during atrial contraction.
- The sino-atrial node initiates the heartbeat by emitting waves of depolarisation across the atria; the atrioventricular node relays the waves to Purkyne fibres so that ventricular contraction starts at the base of the ventricles.
- An electrocardiogram (ECG) shows the electrical activity of the heart. ECGs show the activity of the heart in different conditions, such as fast and slow heart rates and irregular heartbeats.
- Arteries and veins have walls lined with squamous epithelium (endothelium), smooth muscle and elastic tissue. Arteries withstand higher pressure so have thicker walls than veins. Capillaries are lined by endothelium alone so there are short diffusion distances for exchange between blood and tissues.
- Blood is a tissue composed of red and white cells, platelets and plasma; tissue fluid is formed from plasma by pressure filtration and bathes tissues. Pressure filtration is the net effect of blood hydrostatic pressure and its oncotic pressure; excess tissue fluid drains into lymph vessels to form lymph.
- Oxygen and carbon dioxide both bind to haemoglobin, which is an iron-containing protein located in red blood cells. Oxygen binds to the iron-containing haem groups; carbon dioxide binds to amino groups on the molecule.
- Carbonic anhydrase in red blood cells catalyses the reaction between carbon dioxide and water to form hydrogen ions and hydrogencarbonate ions. Hydrogen ions bind to haemoglobin, which acts as a buffer to maintain a constant pH. Hydrogencarbonate ions diffuse out of red blood cells into plasma in exchange for chloride ions that maintain electrochemical neutrality. These changes go into reverse in the lungs.
- The affinity of haemoglobin for oxygen is reduced if the carbon dioxide concentration increases; this helps unloading of oxygen in respiring tissues and is known as the Bohr effect.
- Fetal haemoglobin has a higher affinity for oxygen than adult haemoglobin so that it is fully saturated at the gas exchange surface in the placenta.

■ Transport in plants

Key concepts you must understand

Flowering plants need transport systems to move water and assimilates, such as sucrose and amino acids. Assimilates are so called because they have been made from simple substances, such as water, carbon dioxide and ions (e.g. nitrate ions), which have been taken from the environment and converted into compounds that are incorporated into the plant. Water is absorbed by roots and is required by stems, leaves, flowers and fruits. Sugars are made by leaves but are required by roots, storage organs (such as potato tubers), flowers, fruits and seeds.

The parts of the plants where water and assimilates are *loaded* into the transport systems are called **sources**. The parts where they are *unloaded* are called **sinks**. Transport systems are needed because the distances between sources and sinks are large — even in a small seedling the distance is too great for the movement to occur fast enough by diffusion. These distances are much greater in tall trees such as giant redwoods that grow up to 80 metres or more in height.

Rates of metabolism in plants are much lower than in animals so there is no requirement for a transport system that delivers substances as fast as in animals. Because plants absorb simple substances, such as carbon dioxide, water and ions, from their environment they have very large surface areas — of roots and leaves. Exchange surfaces are therefore all over the body rather than concentrated in organs in different parts of the body as in animals. These exchange surfaces are huge and give most plants a very large surface area-to-volume ratio.

There are two transport systems in plants:
- **xylem tissue** for water and mineral ions
- **phloem tissue** for assimilates, such as sucrose and amino acids

Together these two tissues form the **vascular system** in roots, stems and leaves of flowering plants.

The contents of xylem and phloem move by mass flow. Everything within each 'tube' in these tissues moves in the same direction at the same time.

The movement of water in the xylem depends on **transpiration**, which is the evaporation of water from the leaves and other aerial parts of plants. The source of energy to drive transpiration comes from the Sun. The plant provides a system of channels for water to flow, but does not provide energy for the movement of water in the xylem.

There are many damp surfaces inside leaves from which evaporation occurs, because each cell has its own surface for gas exchange. The consequence is that the air spaces are fully saturated with water vapour. As soon as stomata open to allow carbon dioxide to diffuse into the leaf, water vapour diffuses out. This is why transpiration is an inevitable consequence of the very large surface area for gas exchange inside leaves.

Movement in the phloem is **translocation**. Literally, translocation means from 'place to place' but it is the name given to the mass flow of assimilates dissolved in water that occurs in phloem tissue. This is driven by energy from the plant.

Exam tip

Vascular is a term that refers to the vessels that transport substances in multicellular organisms. The vascular system of mammals is the circulatory system; the term cardiovascular system refers to the heart and vessels.

Exam tip

You may be asked why transpiration is a consequence of gas exchange. See Question 10 in the AS-style paper (p. 83) for suitable answers to this question.

Knowledge check 17

Make a table to compare transpiration and translocation in terms of sources and sinks.

Key facts you must know

Figure 24 shows cross-sections of a leaf, root and stem of a typical flowering plant. You should be able to show on such diagrams where the xylem and phloem are situated. Figure 25 is a photograph of the central area of a root, as seen through the high power of a microscope. You should be able to identify the tissues labelled here.

Xylem tissue consists of:

- vessel elements — dead, empty cells arranged into continuous columns called vessels
- parenchyma cells — living cells found between the vessels

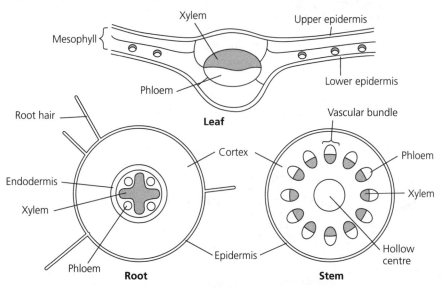

Figure 24 The distribution of xylem and phloem in cross-sections of leaf, root and stem. The separate regions in stems and leaves are vascular bundles

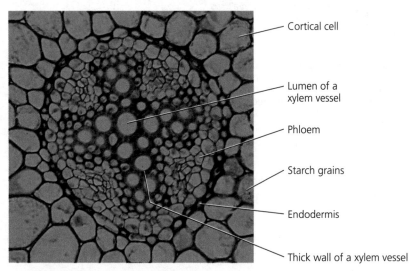

Figure 25 This photomicrograph shows what you should be able to see under the high power of a microscope when you look at the centre of a cross-section of a root. Notice that the xylem in this case is in the form of a cross (×170)

Knowledge check 18

Make a plan drawing to show the distribution of the tissues in Figure 25.

Figure 26 shows details of xylem vessels and how they are adapted to transport water and provide support.

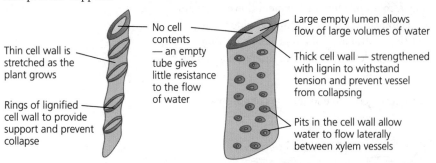

Figure 26 Two xylem vessel elements: (a) a narrow vessel thickened with rings; (b) a wider vessel with pits to allow lateral movement of water

Phloem tissue (Figure 27) consists of:

■ phloem sieve tube elements — living cells arranged into continuous columns called sieve tubes
■ companion cells — smaller, very active cells

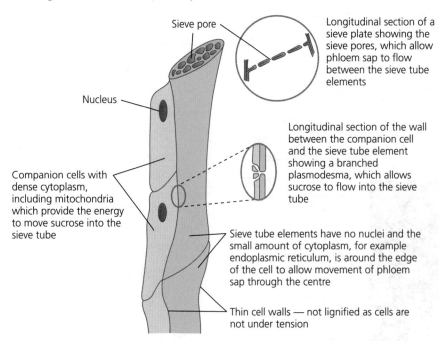

Figure 27 Phloem sieve tubes and companion cells, showing how they are adapted for their functions. Plasmodesmata are small tubes of cytoplasm that pass through the cell wall. They are lined by membrane continuous with the cell surface membrane

Transport of water in plants
Key concepts you must understand

We will start with two important concepts from Module 2 — osmosis and water potential (see the first student guide in this series, p. 53).

Figure 28 shows the movement of water between some mesophyll cells. Cell P has a higher water potential than cells Q and R. This may be because Q and R have more solutes in them than P has, or because they are losing more water by evaporation to the air than P is. The arrows show the direction taken by water. The symbol ψ is used to represent water potential. Remember that −300 kPa is *greater than* −400 kPa, so water moves from a higher water potential (P) to a lower water potential (Q).

Exam tip

Remember that water always moves *down* a water potential gradient. Do not write about water concentrations or concentration gradients for water. You should be able to state the direction water takes when given some water potentials, such as those in Figure 28.

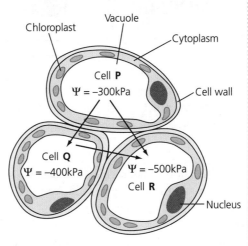

Figure 28 The arrows show the direction of water movement between three mesophyll cells in a leaf. Cell P has the highest water potential and the water potential of cell Q is higher than that of cell R

Key facts you must know

Most of the water absorbed by plants is lost to the atmosphere in transpiration. Water is absorbed by root hair cells and passes across the cortex of the root, through the endodermis and into the xylem in the central region (Figure 29). From here it travels inside xylem vessels until it reaches the leaves, where it may enter cells and:

- be used as a raw material for photosynthesis, or
- enter the vacuole to give it turgidity and help with support, or
- pass to the cell wall and evaporate into the air spaces in the leaf

Water vapour that evaporates from cell walls may diffuse through the stomata into the atmosphere outside the leaf. As water travels across the cortex in the root and across the leaf there are two main pathways that it may follow (Figure 29):

- the **apoplast** pathway — along cell walls
- the **symplast** pathway — from cell to cell through the plasmodesmata

Most of the water probably follows the pathway of least resistance, which is the apoplast pathway. The endodermis is a layer of cells that have impermeable material (the **Casparian strip**) between the cell walls. This is a barrier to the apoplast pathway, so water has to travel through cells into xylem vessels. There is a water potential gradient here, so that water travels from the cortex by osmosis across the

Knowledge check 19

Describe the pathway taken by water as it travels from the soil through a plant to the atmosphere.

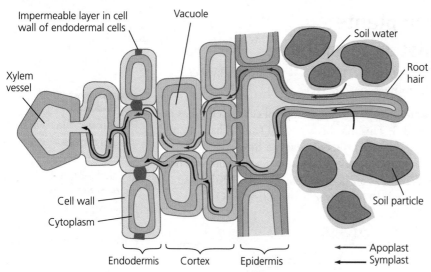

Figure 29 The arrows show the pathways taken by water as it moves from the soil, into a root hair, across the cortex, through the endodermis and into the xylem. Plasmodesmata are shown joining the cells, but they are not as big as shown here. Root hair cells provide a large surface area for the absorption of water

cells and into the xylem. The function of the endodermis is probably to select ions to pump from the cortex into the xylem and then transport to the rest of the plant.

Transpiration stream

The force that draws water up through xylem vessels is **transpiration pull**. Water evaporates from the cell surfaces inside the leaf. This makes the air spaces inside the leaves fully saturated with water vapour. If the stomata are open, then water vapour escapes by diffusion. Transpiration is the combined effect of evaporation from the internal surfaces of leaves and the diffusion of water vapour out of the leaves.

Water moves through plants because of cohesive forces between water molecules and the adhesive forces between water and cell walls. Hydrogen bonds are responsible for both of these. There are hydrogen bonds between water molecules (see the first student guide in this series, p. 15) and also between water and the cellulose fibres in the cell walls of the xylem vessels. The continuous stream of water is maintained by this cohesion–tension mechanism.

Various factors influence the rate at which transpiration occurs:

- Temperature — on hot days, the rate of evaporation increases and the air holds more water vapour. Increasing temperature tends to increase the rate of transpiration.
- Humidity — on very humid days, the atmosphere may hold as much water as the air inside the leaves. This means that there is little or no gradient for the diffusion of water vapour. Increasing humidity tends to decrease the rate of transpiration.
- Wind speed — on windy days, water vapour molecules are blown away from the leaf surface as soon as they pass through the stomata. Increasing air speed tends to increase the rate of transpiration.

Exam tip

Some desert plants open their stomata at night to take in carbon dioxide and then close them during the day to conserve water.

■ Light intensity — in most plants, stomata open during daylight hours to obtain carbon dioxide for photosynthesis. When stomata are open it is inevitable that water vapour will diffuse out of the leaf down the diffusion gradient. At night, plants cannot photosynthesise so they close their stomata to conserve water. As light intensity increases, most plants open their stomata wider to obtain as much carbon dioxide as they can. This tends to increase the rate of transpiration.

A potometer is used to measure rates of water uptake by whole plants, although more often they measure the rate of uptake by cut stems, as in Figure 30. This potometer is used to measure the rate at which a bubble moves along the graduated capillary tubing. A mass potometer is used to measure the rate at which water is lost from whole plants or leafy shoots (Figure 31).

Knowledge check 20

Explain why the volume of water absorbed by a leafy shoot is always slightly more than the volume lost by transpiration.

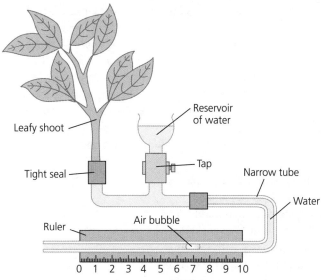

Figure 30 A typical school or college potometer for measuring the *rate of water uptake* by leafy shoots

Figure 31 A simple mass potometer, which measures the loss in mass of the system as a result of transpiration from the leafy shoot. There is a layer of oil on the surface of the water to prevent evaporation

Xerophytes and hydrophytes

Key facts you must know

Xerophytes

Xerophytes are plants that are adapted to living in places where there is a shortage of water. Examples are the cacti that are native to the Americas, although now they are grown all over the world as popular pot plants, and marram grass, *Ammophila arenaria*, which grows on sand dunes that drain very quickly so there is never much water available. There are various adaptations to *reduce* the loss of water vapour, some of which are described in Table 10.

Xerophytes also have adaptations for gaining and storing water. They either have very long roots to obtain water from deep underground or they have very extensive roots that spread out just below the surface of the soil to absorb much water when it rains.

Table 10 Some adaptations of xerophytes to reduce loss of water vapour

Feature	Adaptations for reduction of water loss
Leaves are permanently rolled or roll up in dry conditions, for example marram grass	Air is trapped inside the leaf; water vapour diffuses into the air, but is lost slowly to the atmosphere. The humid air is trapped and reduces the diffusion of water vapour from the stomata. Leaves that do this have their stomata facing inwards when the leaf is rolled.
Thick cuticle	The cuticle is made of waxy substances that waterproof the leaf.
Leaf covered in hairs	Hairs trap a layer of still, humid air; this reduces the rate of diffusion of water vapour from the interior of the leaf.
Stomata sunken in pits or grooves in the leaf	Still, humid air collects in the pits; this reduces the rate of diffusion of water vapour through the stomata

Cacti, like the barrel cactus in the foreground of Figure 32(a), have thick, fleshy stems that are pleated, allowing then to swell when water becomes available. The cactus in the background is covered with many white hairs, which have an albedo effect — reflecting the heat from the Sun. The tall saguaro cacti of the Arizona desert are shaped like cucumbers to reduce the surface area exposed to the Sun at the hottest time of the day.

Exam tip

Never write that xerophytes have 'long roots' without qualifying your answer to say that roots extend deep underground or extend long distances just below the surface of the soil.

Hydrophytes

Hydrophytes are plants that are adapted to live in conditions where there is plenty of water. Some are fully aquatic and live totally submerged, such as Canadian pondweed, *Elodea canadensis*. Others have roots with their stems in water and leaves that float on the surface, like water lilies (Figure 32b). Floating plants have leaves with all their stomata on the upper surface to allow exchange of gases with the air. Oxygen diffuses 100000 times faster in air than in water and carbon dioxide diffuses 10000 times faster, so it is far more efficient to gain these gases from the atmosphere rather than from water. Stems and leaves have large air spaces to provide buoyancy.

Figure 32 (a) Barrel cacti; the species in the foreground is *Ferocactus townsendianus*. Notice the deep folds in the stem. (b) A water lily, *Nymphaea alba*

Exam tip

Expect to find photographs of plants, animals or microorganisms in your exam papers. You are expected to be able to describe their adaptive features (p. 62).

Translocation: source to sink

Key facts you must know

Plants make a great variety of organic compounds. Sucrose is the substance that they make for the transport of energy. Follow the pathway taken by sucrose from **source** to **sink** in Figure 33.

- Mesophyll cells in leaves are the source that make sucrose from the sugars they produce in photosynthesis. Sucrose travels from mesophyll cells (cell A) to companion cells (B), which pump it across their membranes, and then it passes into sieve tube elements through plasmodesmata.
- Accumulation of sucrose lowers the water potential inside the sieve tube elements (C) so that water flows in from surrounding cells by osmosis.
- Hydrostatic pressure builds up inside the phloem sieve tubes and this forces the sugary solution from cell to cell (C to D) through the sieve tubes and away from the leaves. Phloem sap flows from the leaves to meristems (growth areas), roots, new leaves, flowers, fruits and seeds.
- At these sinks, sucrose and other assimilates are removed from the sieve tubes and this lowers the hydrostatic pressure (E).
- This maintains a pressure gradient from source to sink and the mechanism is known as **pressure flow**. Cells in sinks break down sucrose to glucose and fructose, which are metabolised or stored as starch (F).

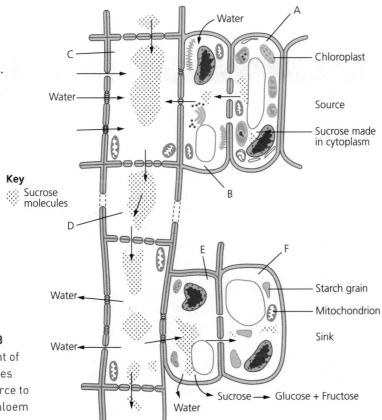

Figure 33
Movement of assimilates from source to sink in phloem sieve tubes

Phloem sap can move in opposite directions in adjacent sieve tubes, unlike the flow of water in xylem, which is always one way — upwards from roots to leaves. Mass flow in phloem is maintained by an active mechanism. There is evidence for this:

■ The rate of flow is higher than can be accounted for by diffusion.
■ Companion cells and sieve tube elements have mitochondria and use ATP to drive pumps to move sucrose. They achieve this by pumping H^+ out of the cell. H^+ diffuses back into the cell through carrier proteins, which also transport sucrose.

Links

It is a good idea to revise the details about osmosis and water potential in Module 2 for this section on transport in plants (see the first student guide in this series, p. 53). You are expected to apply your understanding of these topics to the absorption of water from the soil by root hairs and the transfer of water through a plant. If asked about water, always use the terms 'osmosis' and 'water potential', and explain that water is moving '*down* a water potential gradient'.

The membranes of root hair cells contain aquaporins for absorption of water by osmosis and carrier proteins for the absorption of ions by active transport and by facilitated diffusion (the first student guide in this series, pp. 52–54). The membranes of guard cells have carrier proteins for moving ions into the cytoplasm by active transport. This happens so that the water potential inside the guard cells decreases, water enters by osmosis and the cells become turgid. The swelling of guard cells causes the stomata to open. Companion cells have carrier proteins for moving sucrose.

Exam tip

Remember that movement of water and ions in the xylem and assimilates in the phloem are examples of *mass flow*. Cohesion–tension is the mechanism in xylem; pressure flow is the mechanism in phloem.

Summary

■ Multicellular plants need transport systems because they are large. Distances between roots and leaves are too large for diffusion alone to be effective.
■ Transport (vascular) systems of flowering plants consist of xylem tissue that transports water and ions and phloem tissue that transports assimilates (sucrose and amino acids); the tissues are distributed in the centre of roots and in vascular bundles in stems and leaves.
■ Xylem vessels are columns of dead, empty cells with thick cellulose walls impregnated with lignin; as they are empty there is little resistance to flow and thick walls prevent collapse when water is under tension.
■ Phloem sieve tubes are columns of living cells with little cytoplasm and thin walls. Companion cells use energy from respiration to move sucrose into and out of sieve tubes.

■ Transpiration is the loss of water vapour from aerial surfaces of plants. Water vapour diffuses out down a water potential gradient when stomata are open to allow diffusion of carbon dioxide into leaves.
■ Rates of transpiration increase when it is hot, dry and windy.
■ Potometers measure rates of water uptake into whole plants or into leafy twigs. Most of the water absorbed is lost in transpiration. Mass potometers measure loss in mass of plants or leafy shoots.
■ Water is absorbed by root hairs, diffuses across the cortex to the endodermis through the apoplast pathway in the spaces between cells and in the cell walls. The Casparian strip prevents water passing between cells, so water moves through the symplast pathway into the xylem.
■ Water moves up the xylem by cohesion–tension and evaporates from cell walls of mesophyll cells. This loss of water causes water to move upwards in the transpiration stream.

- The leaves of xerophytes have adaptive features to reduce water loss by transpiration. They also have deep roots and/or shallow extensive roots for the absorption of water. Leaves and stems may be fleshy for the storage of water.
- Water lilies are hydrophytes with leaves adapted to float on water and gain carbon dioxide from the air, not from water.
- Translocation is the movement of assimilates (sucrose and amino acids) from sources (e.g. leaves and storage organs) to sinks (e.g. roots) where they are used.

- Sucrose is loaded actively by companion cells into sieve tube elements. The high concentration of sucrose in sieve tubes in leaves gives a low water potential. Water is absorbed by osmosis, which builds up a high hydrostatic pressure that forces phloem sap from the source to the sink by mass flow.
- Sucrose is removed from phloem in the sink, water follows by osmosis and the hydrostatic pressure decreases. This maintains a pressure gradient between source and sink.

Module 4: Biodiversity, evolution and disease

■ Communicable diseases, disease prevention and the immune system

Communicable diseases and pathogens

Key concepts you must understand

A **parasite** is an organism that gains all or some of its nutrients by living in or on another organism, which is its **host**. A **pathogen** is a parasitic organism that harms its host by causing a **communicable disease**. Table 11 lists several pathogens of flowering plants and mammals (including humans) and the diseases that they cause.

Parasite An organism that lives on or in a host organism from which it gains all or most of its nutrients.

Pathogen A disease-causing organism; all pathogens are parasites.

Communicable disease A disease caused by a pathogen that can be transmitted (communicated) from one host to another.

Table 11 Plant and animal pathogens

Type of pathogen	Disease in plants and *name of pathogen*	Disease in animals and *name of pathogen*
Virus	Mosaic disease (mottled leaves) in many crop plants including tobacco and tomato *Tobacco mosaic virus*	HIV/AIDS in humans *Human immunodeficiency virus (HIV)* Influenza *Influenza virus A, B and C*
Bacteria	Ring rot in potatoes *Clavibacter michiganensis*	Tuberculosis (TB) in humans *Mycobacterium tuberculosis* Bacterial meningitis in humans *Neisseria meningitidis*
Protoctista	Late blight in potatoes and tomatoes *Phytophthora infestans*	Malaria in humans *Plasmodium falciparum*
Fungi	Black sigatoka in bananas *Mycosphaerella fijiensis*	Ringworm in cattle *Trichophyton verrucosum* Athlete's foot in humans *Trichophyton rubrum*

Disease transmission involves the transfer of a pathogen from an infected host to an uninfected host. Pathogens are transmitted directly from one host to another or indirectly. For example:

- **Direct transmission** — the pathogen passes straight from one host to another; examples are by direct contact (mosaic disease of plants), sexual contact (HIV/AIDS) and by droplets in the air (TB). Spores produced by some plant pathogens are transmitted in wind or water (e.g. late potato blight and black sikatoga).
- **Indirect transmission** — another organism known as a vector transfers the pathogen from one host to another. Malaria is an animal example — *Plasmodium* is transmitted by female *Anopheles* mosquitoes; barley yellow dwarf is a plant example — the virus is transmitted by aphids.

Epidemiology is the study of the spread of diseases and the factors that affect it. Living conditions influence the spread of diseases of poverty, such as TB. Weather and climatic conditions influence populations of vectors of plant and animal diseases, which, for example, reach epidemic proportions when conditions favour the reproduction of mosquitoes (malaria) and aphids (barley yellow dwarf).

Plant defences against pathogens

Key facts you must know

Plants have numerous defences against plant diseases. Their surfaces are covered in cuticle and bark, which prevent entry of pathogens. Stomata are a site of entry for many pathogens, but plants can respond to attack by closing them. Once infected, plants respond by making molecules of callose — a polysaccharide — to block sieve tubes and prevent loss of sap. Lignin is also made to 'wall off' infected tissues, depriving the pathogen of any nutrients. The leaves, or other infected parts, then die and fall off the plant.

Host plants also produce a wide range of chemicals, collectively known as **phytoalexins**, which defend against the spread of pathogens. They are **non-specific defences** in that they are released to kill or stop the spread of all types of pathogen. For example, potatoes produce molecules of rishitin that kills pathogens. Plants also have **specific defences** that respond to particular molecules from certain types of pathogens in much the same way that the human immune system responds to human pathogens (see below).

Animal defences against pathogens

Key concepts you must understand

The human defence system consists of primary defences that are physical, chemical and cellular. These prevent entry of pathogens into the body. The second line of defence against pathogens involves chemical and cellular defences that are directed against pathogens once they are inside the body. Phagocytes are part of our **non-specific defence** system. They defend us by destroying invading organisms, but they are not very effective on their own. Lymphocytes and antibodies are part of the **specific defence** system that recognises different pathogens and makes phagocytes more effective.

Disease transmission
The transfer of a pathogen from an infected host to an uninfected host.

Exam tip

Many of the vectors of disease feed on blood or on xylem or phloem. Vectors of plant diseases include aphids, which cause huge economic losses by transmitting diseases of wheat and other crops.

Key facts you must know

Primary defences

The **skin** forms an effective barrier to infection by pathogens. It consists of layers of dead cells filled with keratin, which is a tough fibrous protein. Some pathogens use vectors, such as mosquitoes, to gain entry through the skin.

The gut, airways and reproductive system are lined by **mucous membranes** that consist of epithelial cells interspersed with mucus-secreting cells. Mucus is a slimy substance full of glycoproteins, which have long carbohydrate chains to make them sticky. Small particles in the air, such as bacteria, viruses, dust and pollen, stick to mucus on the lining of the airways; cilia move mucus up the airways to the back of the throat. Coughing and sneezing are **expulsive reflexes** that eject mucus from the nose and throat.

Cells in the lining of the stomach secrete hydrochloric acid that kills bacteria in food. Cells along the length of the gut secrete mucus to protect the lining against attack by acid, enzymes and pathogens. These defences prevent entry of pathogens into the tissues and the blood. They also prevent pathogens growing inside the lungs, gut and reproductive system.

The response of tissues to damage is **inflammation**. Mast cells release the cell-signalling compound histamine to stimulate capillaries to widen to allow an increased blood supply to the area. This brings non-specific cellular and chemical defences such as neutrophils.

Platelets release chemicals that stimulate the **blood clotting** cascade — a series of changes to blood proteins resulting in the conversion of the soluble protein fibrinogen to insoluble fibrin, which forms a meshwork across the wound. Blood cells get caught in the fibres, which dry to form a scab. Below the scab, stem cells divide by mitosis and differentiate into specialised cells of the surrounding area in **wound repair**.

Non-specific defence
Physical, cellular or chemical defence against infection by any type of pathogen.

Specific defence
Cellular and/or chemical defence that acts only on particular types of pathogen or only on particular strains of pathogen.

Cellular defences

Within the body there are two types of cell that are involved in defence:

- phagocytes
- lymphocytes

Both types of cell originate in bone marrow.

Phagocytes

There are two types of phagocyte — neutrophils and monocytes/macrophages.

Neutrophils circulate in the blood and spread into tissues during an infection. They are the 'rapid reaction force' of the immune system, responding quickly by rushing to an infected area and attempting to destroy any pathogens in the tissues. They do not last long. After engulfing bacteria and destroying pathogens, neutrophils die and sometimes accumulate to form pus.

Monocytes pass out of the bloodstream and enter tissues where they form **macrophages** (literally 'big eaters'). They are long-lived cells that have special roles to play in the immune response. Phagocytosis is illustrated in Figure 34. Bacteria are engulfed into **phagosomes**. **Lysosomes** containing hydrolytic enzymes fuse with the phagosomes to form **phagolysosomes**, where bacteria are digested.

Exam tip

Lysosomes contain enzymes such as proteases, carbohydrases and nucleases that catalyse the hydrolysis of proteins, carbohydrates and nucleic acids. They are intracellular enzymes as they work *inside* cells.

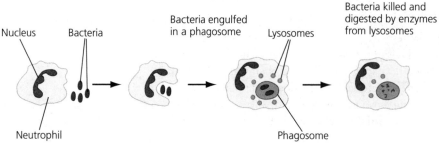

Figure 34 The stages of phagocytosis. Phagolysosomes form when lysosomes fuse with phagosomes to release their hydrolytic enzymes

Lymphocytes

B lymphocytes (**B cells** for short) originate and mature in bone marrow and then spread out through the body's lymphoid system. **T lymphocytes** (**T cells**) originate in bone marrow and migrate to the thymus where they mature. They then spread out through the lymphoid system.

As they mature, B and T lymphocytes gain their own unique **cell surface receptors**. These receptors are glycoproteins and are like antibody molecules. They give the cells the ability to recognise specific antigens (see below). There are small groups of specific B and T cells, each with their own receptors. Although there are many B and T cells in the body, there is only a small number of each type.

Antigens

An **antigen** is a molecule that can stimulate the formation of **antibodies**. Pathogens are covered in molecules (such as proteins and large carbohydrates) that have specific shapes and act as antigens. The immune system recognises any substance foreign to the body as antigenic. These antigens are **non-self** antigens to distinguish them from our own proteins and glycoproteins that are **self-antigens**.

The immune response

Figures 35 and 36 show how lymphocytes respond during an immune response.

B lymphocytes respond to antigens in slightly different ways:

- Antigens that are repeated on cell surfaces, such as polysaccharides on the surfaces of bacteria, are recognised by specific clones of B cells, which divide and start to secrete antibodies, as shown in Figure 35(a).
- Soluble antigens, such as toxins released by bacteria, are recognised by specific T cells, which stimulate specific B cells to respond, as shown in Figure 35(b).

Antigen, antibody and antibiotic are easily confused.

Antigen A compound that stimulates antibody production.

Antibody A protein produced by plasma cells in response to a specific antigen.

Antibiotic A medicinal drug used to treat infections by bacteria.

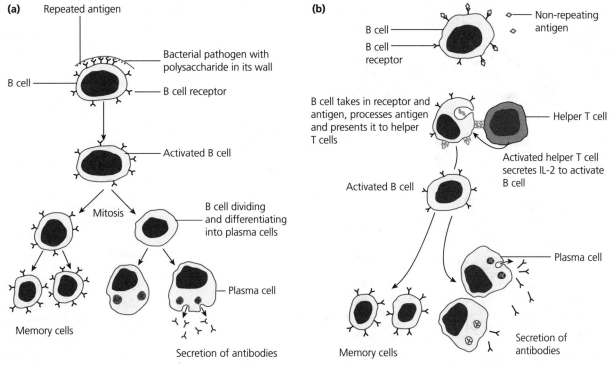

Figure 35 The activation of B lymphocytes by (a) repeated antigens on the surface of cells and (b) soluble, non-repeating antigens

Clonal selection

Of the huge number of small clones of B cells with different receptors in the body, only those that have receptors complementary in shape to the antigens respond by dividing and differentiating into plasma cells. These plasma cells secrete antibody molecules, which are specific to the antigen. Clones of B cells that have receptors that are not specific to the antigen do not respond.

Clonal expansion

T cells are also activated during the immune response and they release cytokines that signal to selected B cells to divide by mitosis. This increases the number of plasma cells so that the concentration of antibody molecules can increase steeply during the immune response.

Antibody secretion

Plasma cells have all the features of protein-secreting cells like those in the pancreas (see the first student guide in this series). Cytoplasm is filled with RER for translation on ribosomes. There is a Golgi apparatus that modifies the protein molecules to become glycoproteins and packages them into vesicles, which travel through the cytoplasm for exocytosis into tissue fluid and blood plasma.

The activation of T cells during an immune response is shown in Figure 36.

Exam tip

This is a good point to revise protein synthesis from Module 2 (Student Guide 1). Plasma cells are good examples of the structure and function of protein-synthesising cells. Recall the difference between transcription and translation as well as the roles of RER, Golgi apparatus, vesicles and the cell surface membrane.

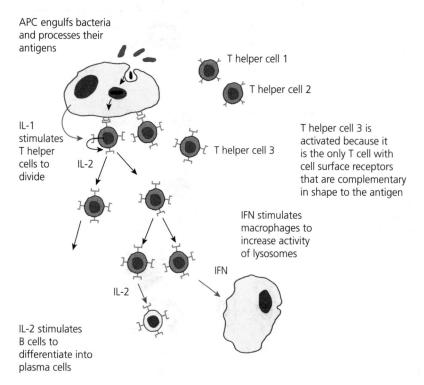

APC engulfs bacteria and processes their antigens

T helper cell 1

T helper cell 2

IL-1 stimulates T helper cells to divide

IL-2

T helper cell 3

T helper cell 3 is activated because it is the only T cell with cell surface receptors that are complementary in shape to the antigen

IFN stimulates macrophages to increase activity of lysosomes

IFN

IL-2

IL-2 stimulates B cells to differentiate into plasma cells

Figure 36 T cells and the immune response

Exam tip

The release of cytokines is an example of cell signalling. Cytokines have shapes that are complementary to those of receptors on cell membranes of B cells, macrophages and other cells of the immune system.

Antigen presentation

Macrophages play an important role in the response to pathogens by engulfing them and partially digesting them. Parts of the molecules from the surface of these pathogens are exposed inside special receptors on the cell surface. T cells respond to this presentation by dividing and secreting **cytokines**, such as **interleukin** 2 (IL-2) to activate B cells to secrete antibodies and interferon (IFN) to activate macrophages to carry out phagocytosis.

Viruses remain in the blood for a short time before they invade our cells. This means that antibodies against viruses do not work well because they are only effective when viruses are in the blood plasma. Antibodies are big molecules that cannot cross cell membranes to combine with viruses inside our cells.

Some bacteria (e.g. *Mycobacterium*) invade cells too. These intracellular pathogens often produce proteins that are expressed in host cell surface membranes, so indicating that the cells are infected. During an immune response, clones of T killer cells may be selected and stimulated to divide by mitosis so that they attack host cells that express these antigens. Activated T killer cells attach to the surface of infected host cells and kill them by releasing toxic chemicals, such as hydrogen peroxide, and enzymes.

The immune response takes several days when an antigen enters the body for the first time — this is the **primary immune response**. This is why we are ill when we catch an infectious disease, such as measles. But after a while, antibodies and activated

Cytokines Small protein molecules that act as cell signalling compounds; many are involved in stimulating responses to infection, such as inflammation and the immune response.

Interleukins Cytokines that are secreted by a variety of cells, especially T helper cells to control the immune response.

T killer cells are produced that help to remove the infectious agent and we recover. The whole process is much more efficient the next time because of memory cells (see below). Immune responses need to be 'shut down' when the pathogen is destroyed. This is the function of **T regulator cells**, which release cytokines to reduce the activity of other lymphocytes.

Memory cells in long-term immunity

Plasma cells do not live long and soon the antibody molecules that they make are broken down in the liver. However, when an antigen enters the body for a second time, the response is much faster. This is because during clonal expansion, B and T cells form memory cells. These remain circulating in the blood and lymph, patrolling the body 'on the lookout' for the return of the same antigen. When this happens, they respond more quickly because there are more of them to be selected than there were of the original clone before the first infection. As a result, the **secondary immune response** occurs much faster than the primary response and there are rarely any symptoms of the infection.

Antibodies: special proteins

It helps here to recall your knowledge of protein structure from Module 2. The simplest form of antibody molecule (known as immunoglobulin G, or IgG) is composed of four polypeptides. Each molecule has two antigen-binding sites that are identical — each binds with the same antigen (Figure 37). This binding is possible because the shape of the binding sites is complementary to that of the antigen. The antibody is specific to the antigen.

Three functional types of antibodies are:

- agglutinins, which cause bacteria to clump ('glue') together, so making it easier for phagocytes to engulf them
- opsonins, which coat bacteria, making it easier for them to be engulfed by phagocytes
- antitoxins, which combine with toxins secreted by bacteria (e.g. tetanus and diphtheria toxins) to render them harmless

We make many antibody molecules with different binding sites to 'fit' around the different antigens that invade us. This is possible because amino acids can be arranged in different sequences to give a range of three-dimensional shapes. Because these binding sites vary, they are also called **variable regions**. The **constant region** is the same for all IgG antibodies and fits into receptors on phagocytes. This helps neutrophils and macrophages to detect pathogens that have been 'labelled' by antibodies for destruction by phagocytosis.

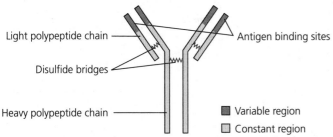

Figure 37 The structure of an antibody molecule (IgG)

Exam tip

T lymphocytes differentiate into the different types (T helper, T killer and T regulator) shortly before and after birth. There are small clones of specific T lymphocytes of all three types scattered throughout the body ready to respond when they detect 'their' antigen.

Exam tip

Figure 37 is a very simple diagram of an antibody molecule. You will get a much better idea of the fit between an antigen and the binding sites of an antibody if you find computer-generated images, a 3D model or watch an animation.

Different types of immunity

So far we have considered what happens when an antigen enters the body. This is **active immunity** (Figure 38). It may happen naturally when you are infected or it may happen artificially when you are given a preparation, or **vaccine**, containing an antigen. Active immunity always involves an immune response.

It is possible to receive antibodies from another person. This is **passive immunity** (Figure 39). There is no contact with the antigen and no immune response occurs. Passive immunity is gained naturally when antibodies cross the placenta or are in breast milk. Injecting antibodies following a snake bite is an example of gaining passive immunity artificially.

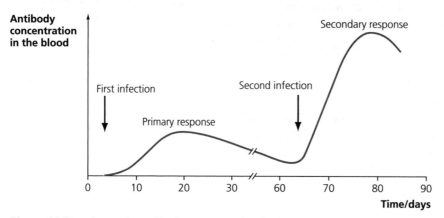

Figure 38 The change in antibody concentration during the primary and secondary response to the same antigen. This is what happens during natural active immunity

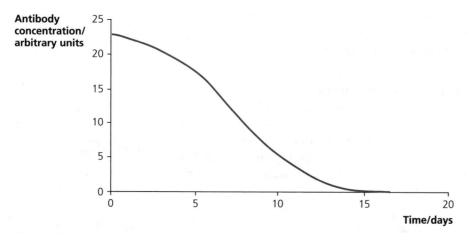

Figure 39 The change in antibody concentration in passive immunity. A person has been injected with antibodies. The concentration of antibodies decreases gradually because they are foreign to the body and are gradually removed. There are no activated lymphocytes (plasma cells) to secrete the antibody

Knowledge check 21

State three ways in which active immunity differs from passive immunity.

Knowledge check 22

Explain the role of memory cells in long-term immunity.

Knowledge check 23

Use Figure 38 to describe how the secondary immune response differs from the primary immune response to the same antigen.

Vaccination controls communicable diseases

Vaccination is **artificial active immunity**. It can be used in two ways:

- **Herd immunity** — as many people as possible are vaccinated so that a pathogen cannot easily be transmitted from an infected person to an uninfected person because everyone, or nearly everyone, is immune. Vaccination programmes attempt to achieve nearly 100% coverage to achieve good herd immunity.

- **Ring immunity** — people living or working near someone infected (or their contacts) are vaccinated to prevent them catching the disease and then spreading it.

People who are vaccinated cannot harbour the pathogen and cannot pass it on to others. This breaks the transmission cycle.

Vaccination programmes are an important part of the health protection offered by governments to their citizens. Programmes for vaccinations against diseases such as polio and influenza are coordinated by the World Health Organization (WHO).

Infants and children are vaccinated against diseases that used to be very common and were responsible for much ill health and many deaths. Many of these diseases are now very rare in most parts of the world. For example, the last case of naturally acquired polio in the UK was in 1984. But the disease still exists in other regions of the world and has been introduced by travellers since then. This is why infants are still vaccinated against this disease as part of their routine immunisation schedule.

The virus that causes influenza changes frequently. The virus can cross-breed with viruses that cause similar diseases in animals or a strain that is pathogenic in animals might cross the species barrier and infect humans. The WHO and national governments maintain a watch for new strains of the virus to which people have not been exposed. Each year, WHO issues guidance about the strains of influenza that are likely to spread. New vaccines are prepared and distributed. People who are at most risk of catching influenza are offered the vaccine.

Autoimmune diseases

The immune system not only mounts defences against pathogens and harmful substances from outside the body, but it can also attack our own tissues leading to symptoms that range from the mild to the debilitating. Diseases of this type are **autoimmune diseases**. They occur because the immune system attacks one or more self-antigens, usually proteins. These attacks often involve antibodies and T killer cells. Table 12 lists some of the more common autoimmune diseases.

Table 12 Examples of autoimmune disease, the parts of the body affected and the main effects of each disease

Autoimmune disease	Area of body affected	Main effects of the disease
Myasthenia gravis (MG)	Neuromuscular junctions	Progressive muscle weakness
Rheumatoid arthritis	Joints	Progressive destruction of the joints
Type 1 diabetes	Islets of Langerhans — endocrine tissue in the pancreas	Destruction of cells that secrete insulin
Systemic lupus erythematosus (lupus)	Skin, kidneys and joints	Progressive deformity

> **Exam tip**
>
> Children are vaccinated against diseases, such as measles, to give herd immunity. Ring vaccination was used in the eradication of smallpox. Transmission of the disease was halted by vaccinating people in areas where cases were identified. Polio is likely to be the next disease to be eradicated.

> **Knowledge check 24**
>
> Explain why the same vaccine for influenza is not used year after year.

> **Exam tip**
>
> MG and type 1 diabetes are good examples to learn as they link to other topics in Module 5. Both systems that are attacked by the immune system are involved in cell signalling — between nerves and muscle (MG) and by secretion of insulin (type 1 diabetes).

The future for medicines

Antibiotics are compounds derived from fungi (e.g. penicillin from *Penicillium*) or from the actinobacteria, particularly *Streptomyces*. This group of bacteria is the source of most antibiotics, such as streptomycin, erythromycin and tetracycline, which are used to treat bacterial infections.

Many antibiotics and other drugs are semi-synthetic in that they are modified chemically after production by microorganisms in fermenters or they are produced entirely by chemical synthesis, although they may have been originally discovered in organisms.

Many pathogens, such as *Mycobacterium tuberculosis*, are now resistant to commonly used antibiotics. Pharmaceutical companies are investing heavily in research to develop new antibiotics and medicinal drugs for other treatments.

There are several ways in which new medicinal drugs are discovered and developed:
- Identifying likely compounds produced by organisms such as fungi, actinobacteria, plants and animals.
- Genetic analysis of organisms to search for likely genes that may code for potential drugs.
- Finding molecules that fit into drug targets, such as receptors for hormones and receptors for neurotransmitters at synapses.
- Modifying existing drugs using computer modelling of the molecular structure of the drug and its target molecule.

Potential sources of new medicines include the following:
- Marine actinobacteria have been discovered to be a source of rifamycin — an antibiotic effective against bacteria as it inhibits protein synthesis.
- *Calophyllum lanigerum*, a rare tree from the rainforest in Malaysia, is the source of calanolide A — a drug that stops HIV entering the nuclei of healthy T lymphocytes. This prevents the T cells producing new viruses and therefore decreases the spread of HIV throughout the body.

Plants used in traditional medicines are likely to make good potential medicines — many drugs in use today are derived from plants. It is likely that animals too may be sources of new drugs. This is one reason why it is important to conserve the world's biodiversity (Table 14 on p. 56).

The technology now exists to sequence people's DNA. The data can be analysed to assess the risks of developing different non-communicable diseases and the likely success of different drug treatments. Drugs work by interacting with target molecules, such as receptor proteins on cell surface membranes. The structure of these receptors is determined by the sequence of bases in the genes that code for them. Analysis of DNA will tell whether individuals have suitable drug 'targets' or not. This is used in **personalised medicine** to decide which drugs should be prescribed when several different ones are available.

Gene technology is also used to modify the genes that control the proteins, such as insulin, that are synthesised in genetically modified bacteria. It may soon be possible to make an artificial prokaryote with the enzymes needed to carry out the synthesis of medicines that are difficult to make in other ways. This fusion of different strands of biology and medicine is known as **synthetic biology**.

Knowledge check 25

Distinguish between an antibiotic and an antibody.

Exam tip

No new class of antibiotic has been introduced since the late 1980s. Antibiotic resistance is a massive global problem. There is more about this on p. 66.

Personalised medicine
Choosing treatments for people based on knowledge of individual personal genetic profiles rather than using the same treatment for all.

Synthetic biology
Genetically modifying organisms to synthesise specific drugs and other chemicals.

Stem cells in bone marrow develop into red and white blood cells. There is more about this in Module 2. Phagocytosis is a form of endocytosis, which you studied in the same module.

Antibodies like the IgG molecule shown in Figure 38 show all four levels of organisation of protein molecules. Remember that quaternary structure is having more than one polypeptide. The 3D shape of protein molecules is crucial here. The shape of the variable region is complementary to the shape of the antigen. This makes an antibody *specific* to its antigen.

Summary

- Parasites live in or on host organisms; parasites that cause disease are pathogens.
- Plant and animal pathogens are viruses, bacteria and protoctists. Disease transmission is the transfer of a pathogen from an infected to an uninfected host. Transmission is either direct from host to host or indirect through the body of a vector.
- Animals and plants have three lines of defence against pathogens. Primary defences prevent entry into the body; secondary defences are non-specific as they are directed at all pathogens that enter. Both plants and animals have specific responses that are directed at certain pathogens; antibodies in animals are an example.
- Phagocytes in blood and tissues ingest and digest bacteria. The immune system provides a highly specific defence against pathogens. During an immune response specific B lymphocytes are activated to secrete antibodies.
- Antibodies are proteins with quaternary structure; the specific shape of the variable regions of each type of antibody is complementary to the antigen to which it binds. Antibodies act as agglutinins, opsonins or antitoxins.
- During immune responses, T lymphocytes are activated to secrete cytokines to signal B lymphocytes to divide by mitosis. During the primary immune response to an antigen, B and T lymphocytes form memory cells that respond faster during secondary immune responses, giving long-term immunity.
- Natural active immunity involves production of antibodies during an immune response to an infection. In natural passive immunity there is no immune response and antibodies are transferred from mother to fetus or baby. Active immunity is permanent; passive immunity is temporary as there is no immune response and no memory cells are produced.
- Vaccination is artificial active immunity that provides long-term immunity and aims to stop transmission of pathogens in populations. Injection of antibodies provides immediate, but short-term, artificial passive immunity.
- The WHO coordinates vaccination programmes to protect people against communicable diseases, such as influenza.
- Personalised medicine involves choosing medicinal drugs according to people's genetic profiles. Synthetic biology is modifying organisms to make medicines and other chemicals that are difficult or expensive to make any other way.

Biodiversity

Key concepts you must understand

At its simplest, biodiversity is a catalogue of all the species in an area, a country or even the whole world. Biodiversity is considered at different levels:

- Ecosystem diversity — how many different ecosystems there are in a specific area (also known as habitat diversity).
- Species diversity — how many species are present in an area.
- Genetic diversity — how much variation there is in the DNA of each species within a particular area.

Much of this section is concerned with how species and genetic biodiversity can be measured.

Key facts you must know

What is a species?

The definition of a **species** is often given as:

A group of organisms able to interbreed and give rise to fertile offspring.

This definition describes a **biospecies**. When many species are described for the first time it is impossible to apply this definition because most are described using physical features, such as morphology (outward appearance) and anatomy. A different definition is used for a **morphological species**:

A group of organisms that share many physical features that distinguish them from other species.

Local biodiversity

A species list gives an indication of the **species richness** of an area — how many species are present. Figure 40 shows students carrying out a survey of the biodiversity of a stream — one of several different habitats that you may study. See Table 13 for their results.

Figure 40 Students carrying out a biodiversity survey

Species A group of organisms able to interbreed and give rise to fertile offspring.

Habitat A place where an individual, population or **community** lives.

Community All the **populations** of organisms living in a well-defined area at the same time.

Population A group of individuals of the same species living in the same area at the same time.

Ecosystem A community and all the abiotic factors that influence it.

Counting the number of species is one way to assess biodiversity. Another is to look at the number of different habitats available in an area. Compare deciduous and coniferous woodland in the UK. An oak tree provides habitats for about 200 different species. Coniferous woodland is dark all year round, has almost no ground flora and the soil has a low pH and few invertebrates. Coniferous woodland provides few habitats, so its biodiversity is low on two counts.

A species list is *qualitative*. It does not give us an idea of the number of each species and so tell us which are common and which are rare. The second approach to assessing biodiversity is therefore to record how many of each species are present — their **abundance**. The number of species and their abundance give another measure of biodiversity — **species evenness**.

Recording the abundance of some plant species is relatively easy. For example, you can count the number of shrubs and trees present in a small area of woodland. Counting the number of birds can be difficult and it is impossible to count the individual grass plants in a lawn; so different methods are employed for different species.

Sampling

Random sampling

Imagine surveying the plants growing on waste ground, many of which are individual plants. It would take too long to count all the plants in the whole area, so we take representative **samples**. A quadrat is used to delimit an area of ground. Most quadrats are $0.5\,m \times 0.5\,m$ or $0.25\,m^2$.

We could do all our sampling at the easiest places to sample, but that would give a biased set of results. The easiest places to sample may have the fewest plant species, so **random samples** must be taken to avoid any bias on the part of the person doing the sampling. This can be done by placing tape measures at right angles to each other along two sides of the sample area. Random numbers are generated using an app to give coordinates where the quadrats can be placed. The numbers of individual plants within the quadrat are counted and the information recorded in a table.

When sampling plant species in other habitats, such as lawns, it is often necessary to record percentage cover because it is impossible to see individual plants. In that case you estimate how much of the quadrat area is taken up by each species and express the answer as a percentage.

Abundance data can be used to calculate **Simpson's index of diversity**. Table 13 shows you the steps involved for the animals found by the students carrying out the stream survey in Figure 40.

The formula is:

$$D = 1 - \Sigma \left(\frac{n}{N}\right)^2$$

The symbol Σ means 'the sum of'.

In this example of the animals in the stream, the index of diversity (*D*) is $1 - 0.3761 = 0.62$ (to 2 decimal places).

Knowledge check 26

Distinguish between species richness and species evenness.

Sampling Taking results from some individuals from a population or small areas within a habitat

Random sampling Each individual or area selected is not chosen by the investigator, to ensure results are representative and unbiased.

Step 1: make a table with four columns

Step 2: write in the species list

Step 3: write in the number of each species

Step 5: divide the number of each species by the total number of individuals

Step 6: square the number in the previous column

Table 13 Data for calculating Simpson's index of diversity

Species	Number in sample (n)	Number of individuals of each species (n) divided by total number of individuals (N)	$(n/N)^2$
Freshwater shrimp	40	0.1544	0.0239
Mayfly nymphs	150	0.5792	0.3354
Water louse	13	0.0502	0.0025
Caddis fly larvae	5	0.0193	0.0004
Lesser water boatman	11	0.0425	0.0018
Water mites	7	0.0270	0.0007
Freshwater annelids	27	0.1042	0.0109
Water snails	6	0.0232	0.0005
Totals	**259**		**0.3761**

Step 4: calculate the total number of individuals

Step 7: calculate the total for $(n/N)^2$

What does this mean? When the index is small (near 0) there is a very low diversity. When the number is high (near 1) there is a very high diversity. Of course, you should realise that this calculation is only made for some of the animals in the stream. The students have only identified some of the invertebrates and have only identified two of them to species level (freshwater shrimp — *Gammarus pulex*) and water louse (*Asellus aquaticus*). This index of diversity should only be used to compare similar habitats and only when identifications are done in the same way.

Non-random sampling

Sometimes it is not possible or appropriate to carry out random sampling. In some places, it is only safe to sample easily accessible places such as either side of paths. Sampling involves trampling over habitats, some of which may be vulnerable, and so access is restricted, so **opportunistic sampling** is carried out where it is safe to do so.

If you wish to find out how biodiversity changes from one area to another **systematic sampling** is used by placing a tape or rope along the ground and taking samples at regular intervals. If the samples are qualitative then this is a **line transect**. If quadrats are used to take quantitative results then this is a **belt transect**.

In some places, there may be two or more distinct habitats. An example is a woodland that is divided by wide paths that are mowed to provide access for horse riders, walkers and cyclists. **Stratified sampling** involves calculating the percentage of the woodland where the dominant vegetation is either trees or grass and taking random samples proportionately within each area — for example, 80% of samples under the trees and 20% on the paths.

Opportunistic sampling
Non-random sampling by taking samples where they are accessible or where it is safe to do so.

Systematic sampling
Taking samples at regular intervals, for example by using transects.

Stratified sampling
Taking samples that are in proportion to the size of the different habitats within an area.

Genetic diversity

Genetic diversity can be assessed by finding within a species the number of:

- different phenotypes — for example, different breeds of domesticated animal, such as dogs, sheep and cattle
- **variants (alleles)** per **gene** — for example, there are four alleles of the ABO blood group gene
- variants of each protein — for example, the number of different versions of an enzyme, as they have differences in amino acid sequences
- different sequences of base pairs in specific coding and non-coding regions of DNA

A gene is a length of DNA that codes for a specific polypeptide. Many genes exist in different forms, known as **gene variants** or **alleles**. The variants or alleles of a specific gene are always present on the same chromosome in exactly the same position. Some genes have no known alleles, some have two; one human gene at the last count has over 4000. However, each individual diploid organism can, at most, have two of these alleles.

All the individuals in a population contribute two alleles to the **gene pool**. Individuals who are homozygous for a gene (e.g. AA) contribute two identical alleles; heterozygous individuals contribute two different alleles (e.g. Aa). Any gene that has a minimum of two alleles that are both present in the gene pool with a frequency greater than 5% is **polymorphic** (literally 'many forms'). Using various techniques, such as electrophoresis, it is possible to study certain genes in a population and assess what proportion of those genes are polymorphic.

Genetic diversity can be determined by finding the proportion of gene loci that are polymorphic (P):

$$P \text{ (proportion of polymorphic gene loci)} = \frac{\text{number of polymorphic gene loci}}{\text{total number of loci studied}}$$

Another measure of genetic diversity is the mean number of alleles per gene (A):

$$A \text{ (mean number of alleles per gene)} = \frac{\text{total number of alleles at each locus}}{\text{total number of loci studied}}$$

Links

Your knowledge of sampling will be applied to fieldwork in Module 6. You will need to know about different sampling methods and how to use them in the field.

Maintaining biodiversity

Key concepts you must understand

As we have seen, biodiversity exists on different levels. You need to understand first what the threats to biodiversity are and why we should act to maintain diversity. Action is taken at a local, national and global scale.

Exam tip

If you are confident about DNA, the genetics you learned at GCSE and the genetic code then read on. If not, stop and read about these topics first.

Gene A length of DNA that codes for a specific polypeptide.

Gene variant (allele) A version of a gene. Alleles of a gene occupy the same position (or locus) on a specific chromosome, but they have different sequences of base pairs (A–T and C–G).

Exam tip

A gene can also be classified as polymorphic if one of its alleles is present in a population at a frequency of 1% in the gene pool. The choice of 5% or 1% depends on the researcher.

Knowledge check 27

In a study of genetic diversity in *Limulus polyphemus* 25 gene loci were studied. The value of P is 0.64. What does this mean?

Key facts you must know

There are many threats to biodiversity, but three important ones are:

■ **human population growth** — more humans means more land needed for housing, agriculture, industry, infrastructure, transport and recreation, so reducing the space available for natural ecosystems

■ **monocultures of crops**, such as wheat, maize and rice — arable agriculture removes natural vegetation and uses fertilisers and pesticides, which damage the wildlife that can survive in these often hostile, uniform environments that offer few ecological niches

■ **climate change** — all aspects of climate are changing more significantly than would occur naturally as a result of human influences, and these affect the distribution and abundance of species, especially small populations (e.g. isolated islands), those already under threat from disease (e.g. amphibians) and those with specific habitat requirements (e.g. alpine plants)

Table 14 shows some of the many reasons why we should maintain biodiversity and populations of species.

Table 14 Reasons for maintaining biodiversity

Reasons for maintaining biodiversity	Examples
Economic	Natural ecosystems provide food and materials for human populations (e.g. fish and timber). They are also important for providing clean water, and in climate regulation, soil formation and treating waste. Old cultivated varieties and wild varieties of crop plants can provide sources of genes and alleles for improving our crop plants in the future. Old breeds of domesticated animals and their wild relatives can be similarly valuable.
Ecological	The loss of species leads to an imbalance in natural communities — for example, loss of keystone species, which play important roles in ecosystems (e.g. by controlling populations of grazers, pollinating flowering plants, and dispersing seeds of forest trees). Loss of top predators, such as lions, tigers and leopards, often leads to an increase in herbivores, overgrazing, land degradation, soil erosion and loss of biodiversity.
Aesthetic	Areas of natural wilderness and managed countryside (e.g. chalk downland) are landscapes appreciated by many as beautiful places.
Medical	Sources of new medicines — see p. 50 for examples.
Ethical	Humans have a duty to conserve natural ecosystems and their biodiversity for future generations.

Ways of maintaining biodiversity

Conservation is defined as keeping and protecting a living and changing environment and its biodiversity. The best way to conserve species is to do so in their habitats. This is conservation *in situ*. Examples are national parks (e.g. The Lake District), nature reserves (e.g. Slapton Ley in Devon), Sites of Special Scientific Interest (e.g. Gouthwaite Reservoir in North Yorkshire) and Marine Conservation Zones (e.g. the waters around Lundy Island in the Bristol Channel).

The **Environmental Stewardship Scheme** (ESS) was introduced in England in 2005 to encourage farmers to manage their land to help maintain biodiversity.

Exam tip

The ESS replaced earlier programmes including the Countryside Stewardship Scheme and Environmentally Sensitive Areas.

Sometimes it is impossible to conserve a species in its natural habitat as that habitat is shrinking or there are so few specimens left in the wild that they must be removed to **botanic gardens** and **zoos** to ensure survival of the species. This is conservation *ex situ*.

Many botanic gardens have **seed banks** where seed is stored. Seeds are collected from the wild, sorted, dried and stored in cold, dry conditions. They are checked at intervals to see if they are still viable. Seed banks ensure a supply of plants for the future and they are stores of genetic biodiversity — an important reserve of genes and alleles for future breeding programmes or to use for genetically modifying plants of economic importance. It is also a store of plant material that may be useful in providing medicines for the future.

Zoos maintain populations of endangered species. Jersey Zoo is involved with captive breeding and reintroduction projects involving tamarins from Brazil. Howletts Wild Animal Park and Port Lympne Reserve in Kent breed lowland gorillas, *Gorilla gorilla*, and are introducing them to reserves in West Africa. Przewalski's horse, *Equus ferus przewalskii*, has been bred very successfully at Whipsnade and Marwell Zoos and animals transferred to Mongolia where this wild horse became extinct 30 years ago.

Zoos cooperate so that breeding programmes generate genetic diversity and ensure that species do not become inbred — a risk when maintaining small populations.

International conservation

The trade in animals for the pet trade and in animal materials, such as ivory, is huge. Much of this is illegal. The **Convention on International Trade in Endangered Species of Wild Fauna and Flora** (**CITES**) is an international treaty that protects animals and plants from various forms of exploitation. Over 30 000 animal and plant species are protected by being placed on one of three Appendices, of which Appendix 1 has the species most at risk of extinction. Some trade is permitted, but only in exceptional circumstances. Appendices 2 and 3 list those species that are less threatened with extinction, but may be so in the future if trade persists.

The **Convention on Biological Diversity** (**CBD**) was signed at the 1992 UN Conference on Environment and Development (UNCED) in Rio de Janeiro (the 'Earth Summit') and ratified in 1993. Countries that are signatories are required to have plans to protect biodiversity, which include writing and implementing Biodiversity Action Plans (BAPs). These act at national and local levels.

In 2013, an environmental audit of a brownfield site destined to be a theme park at Swanscombe, to the east of the Dartford Crossing, found that it is the habitat of a rare jumping spider, *Sitticus distinguendus*. As this species is a priority species on the UKBAP, developers must take steps to conserve it before approval is granted.

Exam tip

You can find out about the work of the Royal Botanic Gardens at Kew and Edinburgh by looking at the conservation sections of their websites.

Exam tip

Each of the zoos mentioned has a website where you can read more about its work in conservation. You can expect questions on species unfamiliar to you; do not panic, the examiner will be asking about general principles of conservation, as outlined here.

Knowledge check 28

What aspects of biodiversity should be considered in any BAP?

Many people are concerned about the environment, but not sure what steps they can take in conservation. This section deals with conservation both at a local and an international level. You can take an interest in both by reading about different aspects of conservation. You can do conservation work in the UK and overseas. You can visit zoos and botanic gardens to see what they achieve in conserving endangered species.

If you wish to be involved in conservation, both in the UK and overseas, have a look at the websites of the British Trust for Conservation Volunteers and Operation Wallacea.

Summary

- A biospecies is a group of interbreeding organisms that produce fertile offspring. A habitat is a place where an organism lives.
- Biodiversity is the number of species in an area and the genetic variation within species. It also includes the variation in habitats and ecosystems in an area.
- Biodiversity in a habitat is assessed by random sampling, often using quadrats.
- Species richness is assessed by identifying all the species in a habitat; species evenness by assessing the abundance of each species in an area in terms of number per unit area or percentage cover.
- Simpson's index of diversity is a measure of biodiversity and uses species richness and species evenness in the calculation. A high value (near 1) indicates an area with high biodiversity.

- Animal and plant species should be conserved for economic, ecological, ethical and aesthetic reasons.
- Animal and plant species are conserved *in situ* in their natural habitats and *ex situ* in zoos and botanical gardens. Conservation is best *in situ* where the complex requirements of each species are provided in its natural habitat. If the habitat is destroyed, or is too small to support viable populations, then removal and captive breeding may be necessary.
- Seed banks maintain collections of seeds of many species including those that are rare. These may be used in the future as sources of variation for breeding and for restoration of habitats.
- International cooperation is a vital part of conservation. CITES and the Rio Convention on Biological Diversity are two such examples.

Classification and evolution

Classification

Key concepts you must understand

To make order out of the variety of life on Earth, we classify organisms into groups. We all do this: plant and animal; edible and inedible; cultivated and wild; dangerous and harmless; cute and ugly. These are simple classification systems.

All living things are believed to have descended from a common ancestor as they share so many important features of structure, biochemistry and inheritance in common.

Scientists have sorted organisms into groups, given names to those groups and given unambiguous names to all the species that have been described. As our knowledge of species changes so do our ideas about how they should be classified, which reflect the way they are related, which in turn is dependent on how they have evolved.

Key facts you must know

In the eighteenth century the Swedish biologist Carl Linnaeus (1707–78) devised the **binomial system** for naming species, which is still used today. Linnaeus gave every known species two names. The first or generic name begins with a capital letter and the second begins with a lower case letter, for example *Bellis perennis* (common daisy). In print these names are italicised but when handwritten they must be underlined. When a binomial has been used once, it may be shortened (e.g. *B. perennis*), so long as it does not cause any confusion.

Classification is the organisation of living things into groups that are arranged in a hierarchy. Linnaeus devised a hierarchical classification system in which large groups were continually subdivided down to the level of the species. Table 15 shows the main taxonomic groups as applied to the classification of three species of mice.

Table 15 The hierarchical classification of three species of mice

Taxonomic rank	Wood mouse	House mouse	Macleay's marsupial mouse
Domain	Eukaryota	Eukaryota	Eukaryota
Kingdom	Animalia	Animalia	Animalia
Phylum	Chordata	Chordata	Chordata
Class	Mammalia	Mammalia	Mammalia
Order	Rodentia	Rodentia	Dasyuromorphia
Family	Muridae	Muridae	Dasyuridae
Genus	*Apodemus*	*Mus*	*Antechinus*
Species	*sylvaticus*	*musculus*	*stuartii*

Antechinus stuartii is a marsupial mouse that lives in Australia. It is carnivorous — unlike the other two species. Superficially it looks like a mouse, but is not closely related at all, as you can see in Table 15.

As biologists began studying the microbial world, they discovered that organisms were built on two basic body plans — prokaryote and eukaryote. They also realised that Linnaeus's two kingdoms (plant and animal) were not sufficient. There was so much diversity within these groups that in 1969 Robert Whittaker (1920–80) devised the five kingdom classification shown in Table 16. (Note that viruses do not fit into this classification — they have their own system.)

During the latter part of the twentieth century scientists gained information on features, such as molecular biology, biochemistry and cell structure. In the 1970s, bacteria were discovered living in extreme environments such as hot springs. These extremophiles, as they are called, share features with prokaryotes and eukaryotes.

In 1990 Carl Woese (1928–2012) introduced the **domain** as a new taxon above the level of the kingdom, giving greater weight to molecular biology (particularly the

Exam tip

The name for each species involves *both* names. The second name should not be used on its own.

Exam tip

Table 15 shows the hierarchical arrangement of the taxonomic ranks beginning with the domain and ending with species. Each of the groups in the rest of the table (e.g. Eukaryota, Rodentia, etc.) is a taxon (plural: taxa).

Table 16 Some of the features used to categorise organisms into the five kingdoms

Features	Kingdoms				
	Prokaryotae (Monera)	Protoctista	Fungi	Plantae	Animalia
Type of body	Mostly unicellular	Unicellular and multicellular	Mycelium composed of hyphae; yeasts are unicellular	Multicellular; not compact	Multicellular; most have a compact body
Nuclei	✗	✓	✓	✓	✓
Cell walls	✓ (made of peptidoglycan)	Present in some species	✓ (made of chitin)	✓ (made of cellulose)	✗
Organelles and fibres (e.g. microtubules)	✗	✓	✓	✓	✓
Type of nutrition	Autotrophic and heterotrophic	Autotrophic and heterotrophic	Heterotrophic	Autotrophic	Heterotrophic
Motility (ability to move themselves)	Some bacteria have flagella	Some protoctists have flagella or cilia	✗	✗	✓ (muscular tissue)
Nervous coordination	✗	✗	✗	✗	✓
Examples	Bacteria and cyanobacteria (blue-greens)	*Amoeba*, *Phytophthora infestans*, algae, slime moulds	Mould fungi (e.g. *Aspergillus*), yeast	Liverworts, mosses, ferns, conifers, flowering plants	Jellyfish, coral, worms, insects, vertebrates

structure of ribosomal RNA) than to other features. The extremophiles were classified in a separate domain — the Archaea — which is at the same taxonomic level as the bacteria and the eukaryotes.

In the past, the only features that biologists could study were those they could see — the external appearance (known as morphology) and the internal structure (anatomy). Classification systems then were based on physical features, often of dead specimens collected on expeditions by naturalists and explorers. Now other evidence is used, such as similarities and differences between the primary structures of proteins and the sequences of bases in genes.

Classification systems reveal the **phylogeny** of taxa because they group together organisms with many shared features. Phylogeny is the evolutionary history of organisms. Not all classification systems reflect phylogeny; some are purely utilitarian — people who write keys for amateur naturalists often first classify flowering plants according to the colour of the flowers and use this feature to make the key. Flower colour is not a feature that reveals phylogenetic links between groups of flowering plants.

Knowledge check 29

Define the terms taxonomic rank, hierarchy, binomial system and phylogeny.

Links

In Module 2 you studied the difference between prokaryotic cells and eukaryotic cells (see the first student guide in this series). The prokaryotic cells you studied were from the Bacteria domain. Archaeans have characteristics in common with both bacteria and eukaryotes, which suggests that they are similar to the very first organisms on Earth.

Evolution

Key concepts you must understand

When people talk about evolution they are usually referring to two distinct ideas:

- The general theory of evolution states that organisms change over time.
- The special theory states that evolution occurs by the process of **natural selection**.

In 1859, Charles Darwin published *On the Origin of Species*, the book that presents the special theory of natural selection with much supporting evidence. Although our ideas about natural selection have changed over the past 150 years, it remains the best explanation of the numerous observations made by scientists about species and how they change.

Key facts you must know

Variation is the sum of the differences between species and within species.

- **Interspecific** variation is the variation *between* species. This is used to identify species and to classify them as discussed in the previous section.
- **Intraspecific** variation is variation *within* species. This variation is the raw material for natural selection and is the variation discussed in this section.

If you look at different features of plants, animals and microorganisms you can distinguish between two forms of intraspecific variation:

- **Discontinuous variation** — distinct forms without any intermediates, for example: human blood groups; red, pink and white flowers in the snapdragon, *Antirrhinum majus*; drug-resistant and drug-susceptible forms of *Mycobacterium tuberculosis*.
- **Continuous variation** — a range between two extremes with no easily identifiable intermediate groups, for example mass and linear measurements of organisms (height of plants, width of leaves, tail length of mice, etc.)

It is sometimes quite difficult to decide whether a feature shows discontinuous or continuous variation. Remember that examples of discontinuous variation are *qualitative* and cannot be given a measurement, whereas examples of continuous variation can be measured and are *quantitative*. When displaying data about continuous variation, it is necessary to divide the range into arbitrary groups and show the data as a histogram. For example, if you measure the tail length of a group of mice of different ages you can divide the results into different groups — 20–29 mm, 30–39 mm, etc. — and then plot them as a histogram. But you have to decide on the number and the lengths to include in each group (e.g. 20–29 mm or 20–24 mm).

Data for discontinuous variation are already grouped into categories (A, B, AB and O blood groups, for example) and are shown as a bar chart with each category represented by a bar separated from others by space on the graph paper.

Features that show discontinuous variation are controlled solely by genes — the environment has no effect. Features that show continuous variation are influenced both by genes and the environment. Consider the factors that determine the body mass of mice: availability of food, environmental temperature and quantity of stored fat. There are some genes that influence body mass; an allele of one of these genes gives rise to obese mice.

Adaptation

For an organism to exist successfully in an environment, it must possess features that help it to survive. Adaptation is the way in which organisms are suited to their environment. This includes their external appearance (morphology), internal structure (anatomy), the function of body systems (physiology), chemistry of cells (biochemistry), behaviour, reproduction and life cycle.

Measuring variation within species is one way in which adaptation can be assessed. Why do adult mice have tail lengths that are all nearly the same rather than a wide range from tiny tails to extremely long tails? Why do some woodland snails have brown shells, while others have pink ones? Why are leaves on the lower branches of trees larger than those at the top? These features are adaptive and studying them can tell us about the ways in which these organisms fit into their environment. You have to be able to outline the structural, behavioural and physiological features of organisms from different taxonomic groups.

So what are the answers to the questions posed above? Mice use their tails for balance and very short and very long tails would not be suitable. Shell colour of woodland snails is related to camouflage. Lower leaves have a larger surface area to absorb sufficient light for photosynthesis because the light intensity at the bottom of trees is lower than at the top.

Table 17 Adaptations of species of plant, animal and microorganism

| Species | Adaptations | | |
	Structural	Behavioural	Physiological
Black mangrove, *Avicennia germinans*, grows along muddy shores in the tropics	Breathing roots that emerge from anaerobic mud to absorb oxygen from the air	Seeds develop on the parent tree, germinate, grow into seedlings and then drop into the mud	Excrete excess salt from glands on leaves
Wood mouse, *Apodemus sylvaticus*	Large ears and eyes for good vision and hearing	Nocturnal, so avoiding some predators	Temperature regulation adapts to changes in seasonal temperatures (e.g. by changing metabolic rate)
Yeast, *Saccharomyces cerevisiae*	Cross-linking between polymers in cell wall gives it rigidity	Under unfavourable conditions, yeast forms spores	When there is no oxygen, respiration is anaerobic

The mammals that migrated to Australasia about 50 million years ago were marsupials and these evolved to fill many niches including subterranean ones. The two species of marsupial mole (Figure 41) live in the deserts of western and northern Australia. Just like the European mole (Figure 42), which is a placental mammal, they burrow through soil and very rarely emerge into daylight.

Some of the adaptations shared by placental and marsupial moles are:

- short and powerful limbs with huge front claws for digging
- no external ears — the openings into the ear canals are just under the fur
- similar fur — the marsupial mole's fur is described as 'silky' and that of the placental mole's as 'velvety'

Exam tip

Marram grass, *Ammophila arenaria*, is another good example that you can use to illustrate adaptations of a plant. See p. 38 for some ideas.

Knowledge check 30

Explain what is meant by an *adaptive feature*.

Figure 41 A marsupial mole, *Notoryctes typhlops*, eating a gecko. The hard area at the front of its head is an adaptation for burrowing

Figure 42 The European mole, *Talpa europaea*, is a placental mammal. This species has the same adaptive feature as the marsupial mole — large front limbs for digging.

■ vestigial eyes — the marsupial mole has tiny, non-functioning eyes; the placental mole has small eyes with vision limited to the detection of light

The similarities between these species, and the golden moles from southern Africa, might make one think they are all descended from a common mole-like ancestor. This is not so; the common ancestor of these animals was nothing like a mole. These features have evolved independently in different taxonomic groups as adaptations to similar ways of life, exploiting the opportunities offered by living underground.

These species are not entirely alike, as you can see in the photographs — the marsupial mole has a hard area on the front of its head, while the European mole has a thin snout. The marsupial mole has a feature unique among marsupials — its neck vertebrae are fused together, which is an adaptation for digging that has not evolved in the other species of mole.

Darwin's observations

Quite independently two Englishmen, Charles Darwin (1809–82) and Alfred Russel Wallace (1823–1913), proposed a theory to account for the evolution of species. They both had travelled extensively — Darwin around the world on the British Navy's survey ship *HMS Beagle* (1831–1836) and Wallace in Amazonia and the East Indies as an explorer and collector of natural history specimens.

Darwin spent about 20 years developing his theory of **natural selection** and writing several drafts of his 'big idea'. A letter from Wallace, containing a concise account of the same theory, prompted him to write a brief scientific paper outlining the theory. This paper was presented to the Linnaean Society in London on 1 July 1858.

In developing the theory of natural selection, Charles Darwin made four observations.

1 All organisms reproduce to give far more offspring than are ever going to survive.

2 Populations of organisms fluctuate, but they do not tend to increase and decrease significantly over time — they remain fairly constant.

3 There is variation among the offspring — they are not exactly like their parents or exactly like each other.

4 Offspring resemble their parents. Features are transmitted from one generation to the next.

> **Natural selection** The survival of individuals with particular features that adapt them to the environment. These individuals have a greater chance of breeding and passing on their alleles than others.

Struggle for survival

There are finite quantities of resources for organisms. Individuals of the same species need the same resources and have similar adaptations for gaining those resources. If there are more organisms than the environment can support there will be competition between them and many will die, so populations remain fairly constant from generation to generation. Some of the environmental factors that control the sizes of populations of heterotrophic organisms are food, water, disease and predation. Populations of autotrophs are controlled by disease and grazing, but also by access to light, carbon dioxide and water.

Individuals that are adapted to gain resources and avoid catching lethal diseases and being eaten are likely to be those that survive and mate to pass on their genes to future generations. This **differential survival** determines which individuals will reproduce more successfully to pass on the *alleles* that have given them the competitive edge over others. These organisms are better adapted than others to the conditions prevailing at the time. They have a **selective advantage**. As a result there is an increase in the proportion of individuals in populations that have the advantageous adaptation and have the allele that codes for it.

Variation is generated by every generation when it breeds to produce offspring. There are different ways in which this variation is generated but **mutation** is the only way in which totally new genetic material is formed.

When the environment is stable, natural selection acts to maintain the features of a species, but if the environment changes then **selection pressures** in the environment change. This means that organisms showing some features that were previously less advantageous are now the ones that compete well, survive and breed. The generation of variation is necessary if this is going to happen and species are to adapt to changing

> **Selection pressure** Any aspect of the environment that influences the survival of individuals, for example predation, disease and competition for food, space and water.

conditions. This shows how variation, adaptation and selection are important components of evolution.

Here are some examples of selection acting on mice:

- Brown mice are visible to predators on a sandy beach where yellow fur is the best colour to have.
- Small, slender mice tend not to survive in cold climates because they have a large surface area-to-volume ratio and lose heat too quickly. Mice in colder regions tend to be larger than those in warmer regions.
- Artificial selection with the house mouse, *M. musculus*, has given a large number of strains that people keep as pets and are used in laboratories. Artificial selection has revealed much more variation than is obvious in wild populations that are exposed to environmental selection pressures.

Evidence for evolution

Key concepts you must understand

Biologists look at remains of organisms from the past and research the environments in which they lived. This line of evidence shows that life on Earth is ancient, and by studying fossils we can follow the history of different groups of organisms. Study of organisms alive today reveals their ancestry or phylogeny. Evidence from biochemistry is helping to build a more detailed history of life on Earth.

Key facts you must know

Evidence from fossils

Fossils are mineralised or otherwise preserved remains or traces (such as footprints) of animals, plants and other organisms. Fossils are found in sedimentary rocks and chemical traces of fossils are detected in metamorphic rocks.

The oldest fossils are those of prokaryotes found in rocks that are 3.5 billion years old. Chemical traces of prokaryotes have been found in rocks even older than this, indicating that life is as old as 3.9 billion years.

Fossils tell us that environments and organisms have changed over millions of years. For example, in the Grand Canyon in Arizona the Colorado River has cut a deep gorge through layers of rock. At the base there are fossils of prokaryotes that are 1250 million years old. Near the top there are fossils of more recent origin, including corals and molluscs that are 250 million years old. In the middle there are fossils of reptiles, amphibians and terrestrial plants.

Biologists in the nineteenth and early twentieth century compared the morphology and anatomy of species to show evolutionary relationships. The pattern of bones in the front limbs in amphibians, reptiles, birds and mammals is basically the same. This indicates that these animals had a common origin.

Evidence from biochemistry

Many biological molecules are the same in all organisms — examples are DNA, RNA, ATP, proteins, phospholipids, polysaccharides and the coenzymes. This argues for a common ancestry for all life on Earth. Analysis of the amino acid sequences of proteins reveals that proteins from closely related organisms are very similar. The primary structure of these proteins is determined by the sequences of bases in DNA. This means that nucleotide sequences in genes from closely related organisms are very similar.

Natural selection in action

Antibiotics became widespread after the introduction of penicillin in the late 1940s. They proved hugely successful in treating bacterial disease such as TB, but very soon antibiotics such as streptomycin became less effective as *Mycobacterium* developed antibiotic resistance. This happened because some bacteria possessed genes that coded for ways to prevent the effect of the antibiotic. For example, penicillin is effective because it prevents the growth of cell walls of some bacteria. Resistant bacteria have enzymes that can break down penicillin.

When antibiotics are used, any resistant bacteria are clearly at an advantage as they are adapted to the new conditions. The susceptible forms die and the resistant bacteria survive and reproduce to pass on their genes to future generations.

Over time, some bacteria have become resistant to many antibiotics. Examples are the pathogens that cause the so-called 'hospital-acquired infections' — methicillin-resistant *Staphylococcus aureus* (MRSA) and *Clostridium difficile* (*C. diff*). Both have caused deaths in hospital patients with suppressed immune systems, such as the elderly and those who have had organ transplants. The numbers of cases of MRSA and *C. diff* in the USA and UK have decreased in recent years as a result of specific strategies, such as implementing strict hygiene controls in hospitals.

Selection has also happened among insect pests that have been sprayed with insecticides. Insects susceptible to insecticides have died, while resistant forms have survived and increased in number.

Exam tip

In these examples, antibiotics and insecticides are selection pressures.

Links

Darwin spent much time studying variation among wild and domesticated animals, such as pigeons, but he did not know how variation was brought about or inherited. At the time he was writing *On the Origin of Species* in England, Gregor Mendel was working out the laws of inheritance in central Europe (see the fourth student guide in this series). At this stage you should use your knowledge of genes and alleles from GCSE to answer questions in the AS papers. If you take the full A-level, you will develop a better understanding of the relationships between how genes function, how they are inherited and how selection operates to favour some individuals and not others.

Summary

- Classification is the organisation of organisms into a hierarchical system from domain to species. Phylogeny is the evolutionary history of a group, for example a species. Closely related species have many features in common, so classification systems often reflect phylogeny.
- Organisms are classified into five kingdoms: Prokaryotae, Protoctista, Fungi, Plantae and Animalia. Each kingdom has a set of features of structure and function, such as methods of nutrition.
- Each species is given a binomial name: the first name is the genus (or generic) name and the second is the specific name.
- Classification systems were originally devised using observable features, such as morphology (external appearance) and anatomy. Other evidence now used includes similarities between primary sequences of proteins and the sequences of bases in genes in different organisms.
- The three-domain system identifies Bacteria, Archaea and Eukaryota as the largest ranks in hierarchical classification, using what are considered to be fundamental features, such as aspects of rRNA.
- Variation refers to the differences between organisms. Interspecific variation is the sum of all differences between species; intraspecific variation is the sum of all variation within a species.
- Continuous variation is variation in a quantitative feature (e.g. height) that exists between two extremes (e.g. tall and short). Discontinuous variation is variation in a qualitative feature that has no numerical value (e.g. blood group).
- Features that show continuous variation are influenced by both genes and the environment. Features that show discontinuous variation are influenced only by genes.
- Organisms have behavioural, physiological and anatomical features that adapt them to their environment.
- Organisms from different taxonomic groups sometimes show similar anatomical features; examples are marsupial moles and placental moles, which are both adapted to a subterranean way of life.
- Charles Darwin and Alfred Russel Wallace both proposed natural selection as the mechanism for evolution.
- Fossil evidence shows that evolution has occurred over time. Evidence from similarities in DNA and molecules such as ATP shows that organisms on Earth have a common origin.
- Variation is the raw material of evolution. In a population that reproduces sexually, each generation has individuals showing variation in many features. Selection acts on individuals; those that survive to reproduce and pass on their alleles are those that are better adapted to the conditions in their environment at the time.
- Humans influence selection indirectly — for example, the use of antibiotics has led to antibiotic-resistance in bacteria and the use of insecticides to insecticide-resistance in crop pest species. The consequences are that we have to develop new chemicals to control bacterial pathogens and the insect pests that destroy crops and lead to food shortages.

Questions & Answers

Exam format

At AS there are two exam papers. Questions in these two papers will be set on any of the topics from Modules 2–4 in the specification. In addition, there will be questions that will test your knowledge and understanding of practical skills from Module 1 and your ability to apply mathematical skills.

If you are taking AS biology, your exams will be as follows:

Paper number	1	2
Paper name	Breadth in biology	Depth in biology
Length of time	1 hour 30 minutes	1 hour 30 minutes
Total marks	70	70
Types of question	20 multiple-choice questions (1 mark each) and structured questions for 50 marks	Structured questions for 70 marks
Synoptic questions	Yes	Yes

At A-level there are three exam papers. Questions in these three papers will be set on any of the topics from Modules 2 to 6 in the specification. In addition, there will be questions that will test your knowledge and understanding of practical skills from Module 1 and your ability to apply mathematical skills.

If you are taking A-level, your exams will be as follows:

Paper number	1	2	3
Paper name	Biological processes	Biological diversity	Unified biology
Length of time	2 hours 15 minutes	2 hours 15 minutes	1 hour 30 minutes
Total marks	100	100	70
Section A	15 multiple-choice questions (1 mark each) and structured questions for 85 marks	15 multiple-choice questions (1 mark each) and structured questions for 85 marks	Structured questions
Synoptic questions	Yes	Yes	The whole paper is synoptic

About this section

This section contains questions similar in style to those you can expect to find in your exam papers. The limited number of questions in this guide means that it is impossible to cover all the topics and all the question styles, but they should give you an indication of what you can expect.

The questions that follow are mostly based on topics in this Student Guide for Modules 3 and 4. The actual examination papers will not be like this — both papers at AS and all three papers at A-level will cover topics from Modules 2–4. To illustrate this, some of the questions in the A-level section test your knowledge of Module 2.

The AS-style questions are on pp. 70–87. The A-level-style questions are on pp. 87–95. If you are not taking the AS papers you can still use the questions for your exam preparation.

The AS-style questions are similar to those you can expect in AS papers 1 and 2. The A-level-style questions are similar to those in A-level papers 1 and 2. Questions in A-level paper 3 are likely to cover topics and skills from all the modules and you have not covered enough at this stage to show you what questions from that paper will be like. Some of the A-level-style questions are set in the context of topics from Modules 5–6, to give you an idea of how you will be expected to use your knowledge of Module 3–4 in the A-level papers. The A-level-style questions (pp. 87–95) have a total mark of 50, not 100, as is the case for A-level Papers 1 and 2 (see table above).

The answers to the 10 multiple-choice questions are on p. 95.

As you read through the answers to the AS style questions, you will find answers from two students. Student A gains full marks for all the questions. This is so that you can see what high-grade answers look like. Student B makes a lot of mistakes — often these are ones that examiners encounter frequently. I will tell you how many marks student B gets for each question. The A-level-style questions only have model answers, similar to those of student A.

Comments

Each question is followed by a brief analysis of what to watch out for when answering the question (icon **e**). Some student responses are then followed by comments. These are preceded by the icon **e** and indicate where credit is due. In the weaker answers, they also point out areas for improvement, specific problems and common errors, such as lack of clarity, weak or non-existent development, irrelevance, misinterpretation of the question and mistaken meanings of terms.

■AS-style questions

Multiple-choice questions

Question 1

The table shows measurements of the surface area and volume of four single-celled organisms. Which organism has the lowest surface area-to-volume ratio?

(1 mark)

Single-celled organism	Radius/μm	Surface area/μm²	Volume/μm³
A	1	12.6	4.2
B	10	1256.7	4188.8
C	100	1.26×10^5	4.19×10^6
D	1000	1.26×10^7	4.19×10^9

Question 2

Carbon dioxide is transported in the blood in a variety of ways. How is carbon dioxide transported in red blood cells?

A carbaminohaemoglobin

B carbonic acid

C carboxyhaemoglobin

D haemoglobinic acid

(1 mark)

Question 3

Which explains why the domain was introduced into classification?

A Some bacteria have features that they share with eukaryotic organisms.

B There are many differences between groups of eukaryotic organisms.

C There were too many kingdoms.

D Viruses are difficult to classify.

(1 mark)

Question 4

If two species belong to the same class, to which rank must they also belong?

A family

B genus

C order

D phylum

(1 mark)

Question 5

Which resource do habitats not provide for crop plants?

A carbon dioxide and oxygen

B food

C space

D water and mineral ions   (1 mark)

ⓔ The answers to these multiple-choice questions are on p. 95, but try them first and then check your answers.

Structured questions

Question 6

A group of students used a spirometer similar to the one shown in Figure 5 on p. 12. A 17-year-old athlete was asked to breathe into the spirometer while sitting down. Figure 1 shows the trace of her breathing. She was then asked to run very fast for a few minutes. Immediately after she stopped exercising, a second trace was made, as shown in Figure 2. The arrows in both figures indicate when she was breathing through the spirometer.

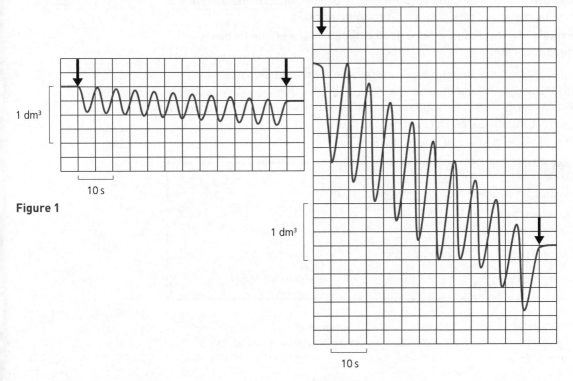

Figure 1

Figure 2

(a) Use the spirometer traces to calculate the following:

   (i)   mean tidal volume (TV) at rest

   (ii) rate of breathing after exercise

In both cases show your working.   (3 marks)

ⓔ To calculate the mean TV you could take measurements of all the breaths, but is that necessary, as it will take a lot of time? Decide how many to use and make the number clear in your working.

(b) Describe how the athlete's breathing after exercise differs from the breathing before exercise as shown by the traces in Figures 1 and 2. Use the data from the two traces to illustrate the points you make in your answer.   (6 marks)

ⓔ This question has 6 marks. You should plan your answer carefully and you can do this by writing notes for yourself on the exam paper. Look carefully to find the differences, list them in your notes and then write them out in full using data taken from the traces. Do not forget to add the units.

(c) Basal metabolic rate (BMR) is the energy released in 24 hours when the body is completely at rest and no food has been eaten for at least 12 hours.

Humans release 21.2 kJ of energy for each 1 dm^3 of oxygen consumed.

   (i) Use this information and the spirometer trace to calculate the athlete's resting metabolic rate in kJh^{-1}. Show your working.   (3 marks)

   (ii) Explain why this result is not a good estimate of the student's basal metabolic rate.   (2 marks)

ⓔ Part (i) is a calculation that involves several steps. The first step involves calculating the oxygen consumption at rest from Figure 1. Questions that test your maths skills in the exam papers will often involve making several decisions rather than just putting numbers into an equation that you are given. There are some clues in the question to help you answer part (ii). Always look for clues.

Student A					
(a) (i) Breaths	1	3	6	9	11
Distance/mm	6	6.5	7	7	6

Mean distance = 6.5 mm

Mean TV = $\dfrac{6.5}{7} \times 500\,cm^3 = 464\,cm^3$

(ii) There were 10 breaths in 62.5 s = $\dfrac{10}{62.5} \times 60 = 9.6$ breaths min^{-1}

ⓔ **3/3 marks awarded** Student A has carefully measured the distance of five of the troughs from the trace. In an exam this would be printed slightly larger so would be easier to do.

Student B

(a) (i) The mean tidal volume = $500\,cm^3$ as each tidal volume is about $500\,cm^3$.

(ii) 10 breaths in $60\,s$ = 10 breaths per minute

ⓔ **0/3 marks awarded** Student B has estimated the mean value without looking carefully at the trace. Many of the troughs are less than $7\,mm$, which is the distance that equates to $500\,cm^3$ on the scale used. Student B has not spotted the position of the start arrow.

Student A

(b) The athlete takes deeper breaths and breathes more slowly. The peaks on the graph fall more steeply, which shows that the oxygen uptake has increased from $250\,cm^3\,min^{-1}$ to $3750\,cm^3\,min^{-1}$. Before exercise her mean tidal volume (TV) was $464\,cm^3$; after exercise it was $2000\,cm^3$. Her ventilation rate (TV × breathing rate) is $20\,dm^3\,min^{-1}$ after exercise; before exercise it was $5.5\,dm^3\,min^{-1}$.

ⓔ **6/6 marks awarded** Student A has written a concise answer using figures to illustrate the points made.

Student B

(b) After exercise the athlete has taken deeper breaths so her tidal volume is much larger. She breathes more slowly than before exercise as the peaks are further apart. This shows that her vital capacity is larger after exercise than before so she breathes in a larger amount of air.

ⓔ **2/6 marks awarded** Student B has not followed the instruction to use the traces and has not measured or calculated anything. This means that marks for use of data cannot be awarded. The answer makes two correct observations about tidal volume and rate of breathing, but the third sentence gains no marks. Figure 2 does not show the vital capacity, which does not change in so short a time — it only increases after a period of training. Student B refers to the *amount* of air. This should be the *volume* of air.

Student A

(c) (i) Over 1 minute the athlete used $250\,cm^3$ oxygen because that is the volume of the decrease in the peaks on the trace:
$$= 0.25 \times 21.2 = 5.3\,kJ\,min^{-1}$$
metabolic rate for an hour = $5.3 \times 60 = 318\,kJ\,h^{-1}$

(ii) This measurement was not taken when she was completely at rest as it was during the daytime (probably during a lesson). It is likely that she has eaten within the 12 hours before the test.

ℯ 5/5 marks awarded Student A has remembered to multiply by 60 to give the answer in kJ *per hour*.

> **Student B**
>
> **(c) (i)** $\dfrac{250}{1000} \times 21.2 = 5.3\,\mathrm{kJ\,h^{-1}}$.
>
> **(ii)** They have recorded the athlete's resting metabolic rate not her basal metabolic rate because she is not completely at rest — she is sitting or standing to breathe into the spirometer.

ℯ 2/5 marks awarded In part (i) student B has used the correct working to calculate the metabolic rate for 1 minute, but has failed to multiply by 60 to give the answer in kJ *per hour*. This means that the unit given is wrong, so no mark is awarded. Student B might do sports studies as the answer refers correctly to *resting* metabolic rate. Full marks are awarded for part (ii).

Question 7

(a) Figure 1 shows three stages in the cardiac cycle.

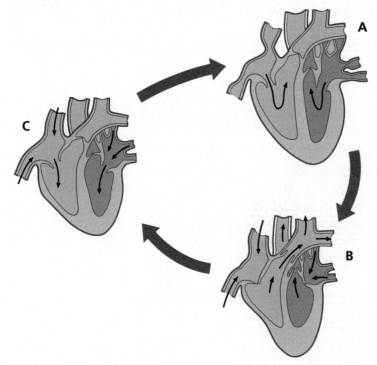

Figure 1

Copy and complete the table at the top of page 75 to show what is happening to the following parts of the *left* side of the heart at each of the stages A, B and C:

- left atrium
- left ventricle
- atrioventricular valve
- aortic valve

(5 marks)

Stage	Left atrium	Left ventricle	Atrioventricular valve	Aortic valve
A	Contracts to force blood into left ventricle	Relaxes and fills with blood from left atrium	Open	Closed
B				
C		Relaxes and fills with blood from left atrium		

ⓔ Make sure that you complete all the cells in the table in questions like this. The examiner has put in plenty of cues in the question to help you to the right answers. Always look for cues.

(b) Figure 2 shows three electrocardiogram traces.

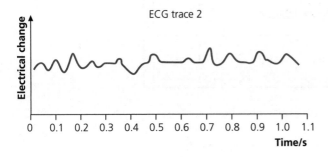

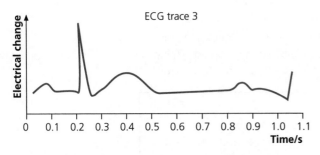

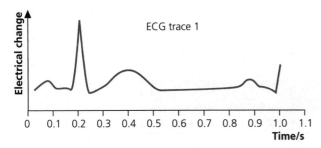

Figure 2

Describe the ways in which ECG traces 2 and 3 differ from ECG trace 1.   (4 marks)

ⓔ Use a ruler here. Move the ruler across the ECGs and mark where significant events occur or, in the case of ECG2, where they do not happen. Using the ruler will help to make precise answers. You are given a timescale on the x-axis, so use it in your answers and do not forget to add the unit — s for seconds.

Questions & Answers

(a)

Stage	Left atrium	Left ventricle	Atrioventricular valve	Aortic valve
A	Contracts to force blood into left ventricle	Relaxes and fills with blood from left atrium	Open	Closed
B	*Relaxes to fill with blood from pulmonary veins*	*Contracts to force blood into aorta (systemic circulation)*	*Closed*	*Open*
C	*Relaxes to fill with blood from pulmonary veins*	Relaxes and fills with blood from left atrium	*Open*	Closed

ℯ **5/5 marks awarded**

(a)

Stage	Left atrium	Left ventricle	Atrioventricular valve	Aortic valve
A	Contracts to force blood into left ventricle	Relaxes and fills with blood from left atrium	Open	Closed
B	*Relaxes to fill with deoxygenated blood*	*Contracts to pump blood out of the heart into the aorta*	*Closed*	*Open*
C	*Contracts to pump blood*	Relaxes and fills with blood from left atrium	*Open*	*Closed*

ℯ **3/5 marks awarded** See Table 5 on p. 22. Student B has not noticed in stages B and C that it is necessary to state where the blood entering the left atrium has come from. The destination for blood pumped out of the left ventricle has been given correctly for 1 mark. A mark is awarded for the state of the valves in B and another mark for the state of valves in C. Re-read pp. 19–21 to make sure you understand what causes atrioventricular valves and semilunar valves to open and close because this often causes problems in answering questions like these. Student B states that the left atrium fills with deoxygenated blood. This is incorrect. Oxygenated blood flows from the pulmonary veins into the left atrium.

(b) ECG2 shows no P, Q, R, S and T waves and there is no regular rhythm. There is very little electrical activity. This is ventricular fibrillation. ECG3 shows a long gap between the P wave and the QRS wave. This could be due to poor conduction in the Purkyne tissue.

ℯ **4/4 marks awarded** Student A gains full marks even though no figures are given. The answer includes correct explanations, even though they are not required by the question. No marks are gained by this extra detail, although it is good to see.

Student B

(b) In ECG2 there is no regular pattern as you can see in ECG1, which means the heart is not contracting properly. ECG3 shows a different pattern of electrical activity from ECG1, with a longer gap before the QRS wave, of about 0.1 s.

ⓔ **3/4 marks awarded** Student B identifies two differences and gains 1 mark for using a figure (0.1 s) from ECG3. Notice how the question is worded: 'describe ways in which ECG traces 2 and 3 *differ from ECG trace 1*. As the question is worded like this, it is acceptable to state features of ECG2 and ECG3 that are not shown by ECG1 without adding 'and these are not in ECG1'.

Question 8

Some students used a quadrat to collect data from a rocky shore. Their results from one quadrat are shown in the table.

Species	Number of individuals in the quadrat	Percentage of the total number of animals surveyed		
Shore crab, *Carcinus maenas*	2	1.1		
Barnacle, *Semibalanus balanoides*	110	58.2		
Dog whelk, *Nucella lapillus*	4	2.1		
Mussel, *Mytilus edulis*	56	29.6		
Common limpet, *Patella vulgata*	17	9.0		
Total	**189**	**100.0**		

(a) (i) Suggest *two* ways in which the results of the survey could be presented so that they could be compared easily with results from other quadrats at different heights on the shore.

(2 marks)

ⓔ Notice that the question says 'two'. It is easy to think of one and write a lot about it and forget that a second answer is needed. Put 1 and 2 on the answer lines to remind you, or use two bullet points.

(ii) Explain why the results are not a good measurement of the biodiversity of the rocky shore visited by the students.

(2 marks)

ⓔ This question tests your skills of evaluation and is asking you to be critical. You need to practise this skill often during your course.

(b) The students took more samples across the rocky shore. Describe a method that they could use to ensure that the quadrats would be placed at random.

(3 marks)

ⓔ This question expects you to recognise the topic of random sampling. You should know a procedure for ensuring that quadrats are placed at random. Throwing them over your shoulder is not one of them.

(c) Calculate Simpson's index of diversity for the results in the table using the formula:

$$D = 1 - \Sigma\left(\frac{n}{N}\right)^2$$

Show your working. (3 marks)

ⓔ This is another multi-step calculation. Look back to p. 54 to see how to set out your answer. It is likely that the table in a question like this will include more columns so that you can simply enter the figures rather than having to write out a complete table.

(d) Suggest why the index for the quadrat on the rocky shore should not be compared with the index for the grassland just above the shoreline. (2 marks)

ⓔ This is a 'suggest' question so there will be a variety of suitable answers, possibly because the examiners do not expect you to know a 'right' answer. Think about the two habitats and see if that prompts some answers.

Student A

(a) (i) The students should use the percentages to plot a pie chart or a bar chart.

(ii) The students took one sample which is unlikely to be representative of the whole shore. The results are taken when the tide is out, more species are found in this area when the tide is in and covers the shore. There are likely to be some species in the quadrat which are too small to see.

ⓔ **4/4 marks awarded** The question refers to data presentation. This is the term in the specification for translation of data from one form into another — here from a table to a pie chart or to a bar chart.

Student B

(a) (i) They could use a kite diagram or several line graphs.

(ii) The problems are that some organisms will be hidden, for example under stones, and they will not find them. Also they only went once — there may have been some animals that only live in this habitat at other times of the year.

ⓔ **2/4 marks awarded** It is not possible to draw kite diagrams from data collected from random samples. Kite diagrams show the results from systematic sampling along a belt transect. Similarly, line graphs are not possible as there is nothing to be used as an x-axis such as distance up the shore. Student B does not gain any marks for part (i), but has written a suitable answer for part (ii).

Student A

(b) The students should make a large square by putting down two measuring tapes at right angles to each other. They should then use random numbers to find coordinates within the large square and place the quadrat in the position within the square as determined by the coordinates. There should be at least 10 quadrats within the area so that means can be calculated.

ⓔ **3/3 marks awarded** Student A has described a generally accepted method for random sampling.

Student B

(b) They could choose a position on the shore and use random number tables to walk a number of paces from that position and then put down the quadrat to take results.

ⓔ **0/3 marks awarded** Student B has the idea of using random numbers, but 'choosing a position on the shore' is not a random method, so this idea cannot gain any marks.

Student A

(c) Student A completed the table as shown below — giving headings to the two columns on the right and filling in the numbers.

Species	Number of individuals in the quadrat	Percentage of the total number of animals surveyed	n/N	$(n/N)^2$
Shore crab, *Carcinus maenas*	2	1.1	0.0106	0.00011
Barnacle, *Semibalanus balanoides*	110	58.2	0.5820	0.33873
Dog whelk, *Nucella lapillus*	4	2.1	0.0212	0.00045
Mussel, *Mytilus edulis*	56	29.6	0.2963	0.08779
Common limpet, *Patella vulgata*	17	9.0	0.0899	0.00809
Total	**189**	**100.0**		0.43518

$D = 1 - 0.43518$

$D = 0.57$ (2 s.f.)

ⓔ **3/3 marks awarded** Student A has completed both columns, which makes the calculations easier to follow and, more importantly, less likely to include any errors.

(c) Student B only completed the final column and finished the calculation by writing:

total = 0.43518

$D = 0.6$ (1 sig fig)

🅮 **3/3 marks awarded** In spite of not completeing both columns, student B has given the correct answer. Answers to one significant figure are accepted here as the lowest number in the table is 2.

(d) They are comparing different types of habitat. The species in the two habitats are very different, for example there are no plants growing on the rocky shore. It may not be possible to identify all the organisms in the field to species level as they have done on the rocky shore.

🅮 **2/2 marks awarded**

(d) The habitats are in two different ecosystems with different geographical and physical features so they can't really be compared using this method.

🅮 **2/2 marks awarded** Both students have given suitable answers for full marks.

Question 9

Pathogens of plants can enter their hosts directly by penetrating the surfaces of leaves. Spores germinate on leaf surfaces to form hyphae that grow into the host tissues. As the hyphae spread inside the leaf, fine extensions, known as haustoria, grow into host cells.

Figures 1 and 2 show two ways of entry into leaf tissues.

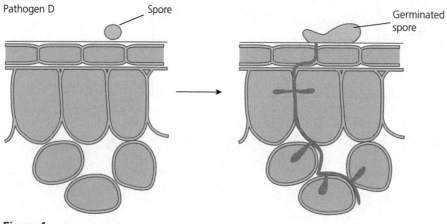

Figure 1

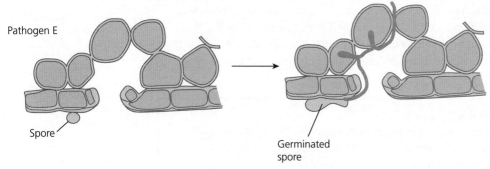
Pathogen E

Spore

Germinated spore

Figure 2

(a) Pathogen D causes the disease late blight and pathogen E causes black sikatoga. State a crop plant that is a host of D and a crop plant that is a host of E. (2 marks)

ⓔ You may never have thought about plant diseases before. The disease caused by late blight altered the course of history in parts of Europe (e.g. in Ireland), and black sikatoga threatens the livelihood of many farmers across the world. Plant diseases are of great economic and social importance.

(b) Describe the differences between the methods of entry of the two pathogens as shown in Figures 1 and 2. (3 marks)

ⓔ This is a question that tests your powers of observation and knowledge of leaf structure. The examiners will look for use of the correct terms for the cells and tissues visible in Figures 1 and 2.

(c) Suggest and explain how crop plants protect themselves against infection by pathogens. (3 marks)

ⓔ There are some clues to answering this question in Figures 1 and 2, if you think of ways to prevent the entry of hyphae into the leaf.

(d) Many viral pathogens of plants are transmitted indirectly. Explain how this is achieved. (2 marks)

ⓔ Notice that the first sentence of the stem of the question uses the word 'directly'. You now have to think of methods that involve transmission that is *indirect*.

Student A

(a) D potato, E banana

ⓔ **2/2 marks awarded**

Student B

(a) D tomato, E oranges

1/2 marks awarded Black sigatoka is a disease of bananas, not citrus trees such as orange. The most important things to learn in this section are the types of pathogen and the diseases that they cause. However, you should also know that many plant pathogens are specific to their hosts and only infect one or several related species.

Student A

(b) Hyphae from D enter directly through the epidermis. The hyphae grow between the cells of the upper epidermis and palisade mesophyll. Hyphae from E grow through the stoma into the intercellular air spaces in the spongy mesophyll. Both D and E grow into the cells.

ⓔ **3/3 marks awarded**

Student B

(b) A grows through the cuticle on the upper layer of cells. The hyphae then grow into the cells. B grows into spaces inside the leaf.

ⓔ **1/3 marks awarded** The only term that student B has used is cuticle and so only 1 mark is awarded. It is correct that A has grown through the cuticle, which was overlooked by student A.

Student A

(c) Make callose and lignin to surround infected areas so hyphae cannot spread. Potato plants produce rishitin, a phytoalexin that inhibits growth of *Phytophthora infestans*, which is the pathogen that causes late blight.

ⓔ **3/3 marks awarded**

Student B

(c) Plants can make thicker cuticle and close their stomata when under attack by hyphae.

ⓔ **2/3 marks awarded** Student B gives two correct points, but there are 3 marks available.

Student A

(d) By vectors. Aphids are pests of plants: they feed by inserting stylets into phloem sieve tubes and sucking the sap. They can take plant viruses into their bodies and transfer them in their saliva when they feed — in the same way that *Anopheles* mosquitoes transmit the malarial pathogen.

ⓔ **2/2 marks awarded**

@ **0/2 marks awarded** Student B has not described a method of indirect transmission, so does not gain any marks.

Question 10

(a) Explain why transpiration is the consequence of gas exchange. (3 marks)

@ Gas exchange occurs on the surface of all the mesophyll cells — it is a huge surface area from which water evaporates.

(b) A student investigated the rate of uptake of water by leafy shoots of two species of terrestrial plant, P and Q, that live in different habitats. The results are shown in the table.

Time/h	Rate of uptake of water/$cm^3 h^{-1}$	
	Species P	Species Q
06.00	2.5	2.3
08.00	5.0	2.0
10.00	25.0	0.5
12.00	36.0	0.2
14.00	42.0	1.0
16.00	45.0	1.0
18.00	49.0	3.5
20.00	36.0	3.8
22.00	10.0	3.7
24.00	7.0	3.9

Explain the difference in the results for the two species. (4 marks)

@ More cues here — the examiner states that the two species come from different habitats and that they are terrestrial and therefore not aquatic. Look at the obvious difference between the figures and do not miss the times of day — read down the columns and then across the rows looking for pairwise comparisons.

(c) Explain the mechanisms that bring about the transport of water in xylem vessels. (5 marks)

@ Always start answers about transpiration stream at the *top* of the plant in the *leaves*.

@ **3/3 marks awarded**

Questions & Answers

> ### Student B
>
> **(a)** Transpiration is the loss of water by evaporation from the mesophyll inside the leaf and diffusion of water vapour through the stomata. Gases diffuse in and out through the stomata.

🅮 **0/3 marks awarded** Student B fails to explain that the damp surfaces inside the leaf are the site of gas exchange, instead suggesting that it occurs through the stomata. This is not so. Gases diffuse through the stomata but the exchange occurs on the cell walls in much the same way as gas exchange in mammals occurs in alveoli (p. 10). These gas exchange surfaces have two features in common: they both form a large surface area and they are thin.

> ### Student A
>
> **(b)** Species P has a much higher rate of water uptake than species Q and is therefore losing much more water by transpiration. Most of the water absorbed is lost as water vapour in transpiration from leaves. P could live in a moist habitat and Q in a very dry habitat, so Q is a xerophyte. The rate of water loss falls in P at night because the plant closes its stomata. Q however has greatest loss in the night and almost none at all at midday. This is because it closes its stomata during the day to conserve water and absorbs carbon dioxide at night through open stomata and then stores it for photosynthesis during the day even though stomata are shut.

🅮 **4/4 marks awarded** Student A gives a very thorough explanation, showing a good knowledge of plants in dry habitats. Plants that only open stomata at night have a special form of metabolism; many are succulent plants that also store lots of water — two adaptive features for desert habitats.

> ### Student B
>
> **(b)** The rate of water uptake of species P is $2.5\,cm^3\,h^{-1}$ at 06.00, increasing to $49\,cm^3\,h^{-1}$ at 18.00. Q is lower at $2.3\,cm^3\,h^{-1}$ at 06.00 and $3.5\,cm^3\,h^{-1}$ at 18.00.

🅮 **0/4 marks awarded** Student B has given data quotes, but has not described any trends or patterns or offered any explanation. This is quite a common error and scores no marks. If a question asks for an explanation it may be necessary to give some description first, as student A has done.

> ### Student A
>
> **(c)** Water is pulled up xylem vessels due to the evaporation of water from the damp cell walls of the mesophyll cells and the diffusion of water vapour to the atmosphere. The transpiration pull along the length of the vessels depends on two types of force: cohesion — a force of attraction between water molecules; and adhesion — a force of attraction between two unlike molecules, in this case water molecules and the cellulose cell walls of the vessels. This cohesion–tension allows net movement of water in an unbroken, continuous column known as mass flow.

ⓔ **5/5 marks awarded** When explaining how water moves through the xylem always start your explanation with transpiration in the leaves and refer to **transpiration pull**, as student A has done here. Student A has also included appropriate terms.

> **Student B**
>
> **(c)** Root hair cells absorb water by osmosis. Water travels across the cortex of the root through the apoplast and symplast pathways until it reaches the endodermis, where there is the Casparian strip that prevents water travelling by the apoplast pathway. The endodermal cells in the root pump solutes and ions into the bottom of the xylem vessel and make a water potential gradient, so water moves into the xylem vessels by osmosis.

ⓔ **0/5 marks awarded** Student B has described movement of water into the root and across the root cortex. This is not transport *in the xylem*, so cannot gain any marks.

Question 11

(a) Explain what is meant by the term *phylogeny*. (1 mark)

ⓔ Remember to learn definitions of all the terms used in the learning outcomes in the specification. The key terms in these Student Guides will help you.

The technique of DNA barcoding is used as a way to distinguish between different species of plants, animals and microorganisms. Barcoding involves sequencing the DNA from certain genes in groups of organisms. It has provided evidence to divide giraffes into eleven genetically distinct groups and helped to separate species of butterfly that appear very similar, but do not reproduce with each other. Studies of amino acid sequences in proteins, such as cytochrome c and fibrinogen, are also used to classify organisms.

(b) List three features that all organisms classified into the kingdom Animalia have in common, other than being eukaryotic. (3 marks)

ⓔ Do not write down the first three things that come into your head. Always think about your answers before you write them down. Often the first things we think about are incorrect or too vague and not specific enough.

(c) Explain the advantages of using evidence from DNA and proteins in research work in taxonomy. (4 marks)

ⓔ Re-read the passage to see if there are any clues or cues to help you. Do not simply repeat the information, but use it to support your arguments.

> **Student A**
>
> **(a)** Phylogeny is the evolutionary history of a species.

ⓔ **1/1 mark awarded** Phylogeny is a term used in the specification and you should be able to define it. Student A has given a crisp definition.

Questions & Answers

Student B

(a) Phylogeny is how you classify species into phyla.

e **0/1 mark awarded** A good idea is to make your own glossary of terms and then learn it thoroughly before the examination. Student B has obviously not done this. Notice how easy it is to see a word and think of the wrong definition.

Student A

(b) Multicellular. Heterotrophic. All animals have a nervous system.

e **3/3 marks awarded**

Student B

(b) Animals move around. Their cells have nuclei. Cells do not have cells walls. They have many cells in their bodies.

e **1/3 marks awarded** Student B has given four answers. The examiner will probably only mark the first three. This is unfortunate, because the student has included the meaning of the term eukaryotic ('their cells have nuclei'), so that does not count. This means that the final answer ('many cells'), which is correct, does not gain a mark. 'Animals move around' is too vague; many animals are sessile (attached to a substrate) and can only move parts of their bodies.

Student A

(c) It can be difficult to tell closely related species apart, especially if you cannot breed them together to see if they produce fertile offspring. Often this is impossible because there are not enough specimens, the specimens have been collected in the wild and are now dead, the specimens have been preserved for a long time or they are fossil species. Over time, the sequence of bases (A, C, G and T) in genes changes. Often this has no effect on the sequence of amino acids in proteins. By comparing the sequence of nucleotides and sequence of amino acids between two different species, you can tell how closely related they are.

e **4/4 marks awarded**

Student B

(c) Scientists have compared the amino acid sequences (primary structures) of proteins like haemoglobin, fibrinogen and cytochrome c. Animals that are closely related (e.g. chimpanzees and humans) show no or very few differences. Animals that are less closely related have many more differences. This information is helpful to confirm classifications based on other features. But DNA sequences are more varied than primary sequences as most amino acids are coded for by more than one triplet. So there are more differences in DNA sequences and by comparing the sequence of nucleotides you get even more data about differences and similarities between species.

ⓔ **4/4 marks awarded** This is quite a difficult topic for you to appreciate in full at this stage. The differences between the DNA of different species are due to mutation, which is a topic covered in Module 5. However, it is appropriate for now to consider the sequences of bases in DNA and the sequences of amino acids in proteins from Module 2. Both students have explained this well, with student B giving some examples of proteins that have been sequenced and used in taxonomic studies. DNA barcoding of animals uses a gene from mitochondrial DNA.

ⓔ Overall, student B gains 26 marks, which may not quite be enough for a pass grade. Marks were lost for a number of reasons:

- Not answering the question as set (e.g. Q.10b).
- Not developing an answer fully enough (e.g. Q.6b).
- Not reading carefully the information provided (e.g. Q.6a).
- Not knowing ways to present data (e.g. Q.8ai).
- Not knowing a practical procedure (e.g. Q.8b).
- Writing more than the number of answers required (e.g. Q.11b).
- Not giving suitable definitions (e.g. Q.11a).
- Not using data provided in graphs or tables to support an answer (e.g. Q.6b).
- Not looking for trends and patterns in data (e.g. Q.10b).
- Not using appropriate terms (e.g. for the leaf in Q.9b).
- In Q.10 it is clear that student B has neglected to learn the plant biology effectively. Plants are important — we would not be here without them!

■A-level-style questions

Multiple-choice questions

Question 1

Which is most likely to contribute to the evolution of an antibiotic-resistant strain of a species of a bacterium?

A the infrequent use of low doses of the antibiotic

B the prescription of the antibiotic for viral diseases

C the use of the antibiotic in combination with others

D the use of the antibiotic to treat specific strains of the bacterium   (1 mark)

Question 2

The role of T helper lymphocytes is to:

A attach to virally infected cells and kill them

B inhibit antibody production by plasma cells

C produce antibodies in response to soluble antigens

D stimulate B lymphocytes to become plasma cells   (1 mark)

Question 3

Which feature of a population indicates high genetic diversity?

A codominance at many loci

B diploid number is in excess of 20

C many gene loci

D many polymorphic gene loci   (1 mark)

Question 4

The rate of flow in phloem is calculated from measurements of:

- the increase in dry mass of a plant organ expressed in terms of the cross-sectional area of the phloem tissue entering the organ
- the concentration of sucrose in phloem tissue

The formula for calculating the rate of flow in phloem tissue is:

$$\text{rate of flow (in } m\,s^{-1}) = \frac{\text{increase in dry mass (in } g\,m^{-2}\,s^{-1})}{\text{concentration (in } g\,m^{-3})}$$

Measurements were made from growing fruits:

increase in dry mass = $10.00\,g\,m^{-2}\,s^{-1}$

concentration of sucrose in phloem = $17.86 \times 10^3\,g\,m^{-3}$

Which is the rate of flow?

A $5.6 \times 10^{-7}\,m\,s^{-1}$

B $5.6 \times 10^{-6}\,m\,s^{-1}$

C $5.6 \times 10^{-4}\,m\,s^{-1}$

D $5.6 \times 10^{-3}\,m\,s^{-1}$ (1 mark)

Question 5

Cardiac output (CO) in $dm^3\,min^{-1}$ is calculated from measurements of stroke volume (SV) and heart rate (HR). Which formula is used to calculate cardiac output?

A $CO = \dfrac{HR}{SV}$

B $CO = HR \times SV$

C $CO = \dfrac{SV}{HR}$

D $CO = (60 \times SV)/HR$ (1 mark)

Structured questions

Question 6

Lysozyme is an antimicrobial agent that is present throughout the body. Lysozyme catalyses the hydrolysis of glycosidic bonds in certain polysaccharides that are present in the cell walls of some bacteria.

(a) State *three* places in the human body where lysozyme is found. (3 marks)

ⓔ There is a clue to one of these places in the name of the enzyme.

(b) Describe what happens when lysozyme catalyses the breakdown of a glycosidic bond in a cell wall polysaccharide of a bacterium. (4 marks)

ⓔ This is an example of a synoptic question because you must use knowledge of enzyme action from Module 2.

Researchers isolated lysozyme from mice to find out how effective the enzyme was at destroying two species of bacteria, *Escherichia coli* and *Staphylococcus aureus*. The researchers prepared five different concentrations of lysozyme. The two species of bacteria were incubated separately in each concentration for 3 hours at 37°C. At the end of the incubation period, the researchers determined the number of bacteria still alive in each concentration of lysozyme and expressed the numbers as percentages of the number of bacteria present at the start of the incubation.

ⓔ Read passages like this at least twice, underlining or circling key information and asking yourself 'why am I being told this?'. For example, why '3 hours at 37°C'?

Concentration of lysozyme/pmol dm^{-3}	Percentage of bacteria still alive	
	E. coli	*S. aureus*
0	100	100
10	65	98
40	40	94
80	15	85
150	0	40

pmol = picomole = 10^{-12} mole

(c) Explain why the researchers included the lysozyme solution of 0 pmol dm^{-3}. (1 mark)

ⓔ The key word here is *explain*. You have to give a reason for including this control. Just writing 'as a control' is not enough at AS or A-level.

(d) Using the information in the table, compare the effect of the different concentrations of lysozyme on *E. coli* and *S. aureus*. (5 marks)

ⓔ When analysing and interpreting data in tables look down the columns and also across the rows. In this question you can write notes about similarities and differences between the figures for the two bacteria at the base of the columns and at the end of the rows. Do not forget to quote figures with units.

(e) Suggest a possible explanation for the different effects of lysozyme on *E. coli* and *S. aureus*. (2 marks)

ⓔ Planning an answer to a question like this should involve thinking about your knowledge of enzymes and prokaryotic cell structure and looking for clues in the whole question (e.g. why *certain* polysaccharides?) and in your other answers.

Questions & Answers

(a) tears, saliva, lysosomes

ⓔ **3/3 marks awarded** Lysozyme is also found, for example, in milk, sweat, mucus and phagolysosomes (for intracellular digestion in phagocytes).

(b) Part of the polysaccharide either side of a glycosidic bond fits into the active site of lysozyme to form an enzyme–substrate complex. Water enters the active site to break the bond. In the reaction hydrogen is transferred to the oxygen of the glycosidic bond. The products leave the active site.

ⓔ **4/4 marks awarded** The answer refers to what happens during the reaction using key terms from Module 2 such as active site and enzyme–substrate complex.

(c) As a control to find out how many bacteria survive without lysozyme when kept under exactly the same conditions.

ⓔ **1/1 mark awarded** The important phrase here is the last three words — the control is to rule out the effects of the conditions under which the experiment is carried out.

(d) All the bacteria survive for 3 hours when no lysozyme is present. As the concentration of lysozyme increases the percentage of bacteria surviving decreases for both bacteria. Lysozyme is more effective at killing *E. coli* as all were killed at 150 pmol dm^{-3}, but 40% of *S. aureus* survived at this concentration. Lysozyme was very effective against *E. coli* at low concentrations, for example 35% died at 10 pmol dm^{-3}, but lysozyme had little effect on *S. aureus* until after 80 pmol dm^{-3}.

ⓔ **5/5 marks awarded** Note the effective use of the figures (with units) taken from the table. Sketching a graph sometimes helps to identify trends and patterns in tables.

(e) *S. aureus* has a thicker cell wall than *E. coli* and it has a polysaccharide with few bonds of the type that are broken by lysozyme.

ⓔ **2/2 marks awarded** In questions of this type it is important to make comparative statements so that differences are relative. Just saying 'a thick cell wall' might not be enough for a mark.

Question 7

In mammals, fetal red blood cells formed in the liver are nucleated and contain fetal haemoglobin. Fetal haemoglobin is a conjugated protein with four haem groups linked to two α (alpha) and two γ (gamma) polypeptides. Fetal haemoglobin is replaced by adult haemoglobin so that by 6 months after birth the blood is composed entirely of adult haemoglobin.

(a) (i) How does the structure of adult red blood cells differ from the structure of fetal red blood cells? (1 mark)

(ii) How does the structure of adult haemoglobin differ from the structure of fetal haemoglobin? (1 mark)

🄔 Read the question carefully for clues to both parts of this question.

(b) Suggest what happens to the genetic material of blood-forming cells in order to change from fetal haemoglobin production to adult haemoglobin production. (2 marks)

🄔 The cue is 'genetic material'. This is set in the context of Module 6 and 'gene switching', but you can probably work out what must happen. Think about the structure of haemoglobin first.

A sample of fetal blood was exposed to increasing partial pressures of oxygen in a chamber with a pH of 7.4. The oxygen concentration of the blood sample was monitored constantly. The procedure was repeated with a sample of maternal (adult) blood. The results are shown in the graph.

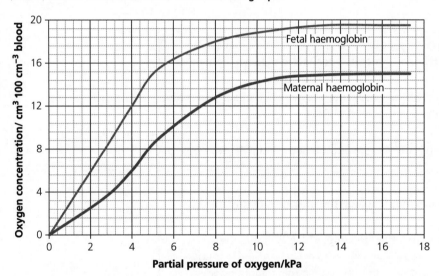

(c) (i) The partial pressure of oxygen in placental tissue is 4.0 kPa. State the volume of oxygen transported by 100 cm³ of fetal blood and adult blood at this partial pressure of oxygen. (1 mark)

🄔 Use a ruler to find the intercepts and read from the graph accurately. Examiners may not allow any leeway in the answer.

(ii) Explain the advantage of the difference between the two types of blood. (3 marks)

🄔 The placenta is an organ specialised for exchange between maternal and fetal blood. That's the only fact you need to know to work out the rest from the graph and your knowledge of dissociation curves.

(iii) Some people are born with an inherited condition in which they are unable to produce adult haemoglobin. Explain why this is a serious condition. (2 marks)

ⓔ Use a ruler to move from right to left along the x-axis and see what happens to the volume of oxygen transported by the two types of blood.

(d) Many ions are of biological importance. Explain the roles of zinc ions and chloride ions in the transport of carbon dioxide in the blood. (5 marks)

ⓔ To gain full marks you must write about both ions. This is where making links between ideas during your revision will prove useful. The link here is carbon dioxide and carbonic anhydrase.

Student answers

(a) (i) Adult red blood cells do not have a nucleus.

(ii) Each molecule of adult haemoglobin is composed of two α globins and two β globins.

ⓔ **2/2 marks awarded** The globins are polypeptides and that would be a suitable alternative to use in (ii).

(b) The transcription of the gene for γ globin stops, so no more is made. The gene is 'switched off'. The gene for β globin is transcribed ('switched on').

ⓔ **2/2 marks awarded** It is much better to say that the gene is transcribed than 'copied'. You could also refer to the production of mRNA.

(c) (i) Fetal blood = $12\,cm^3\,O_2\,100\,cm^{-3}$ blood; adult blood = $6\,cm^3\,O_2\,100\,cm^{-3}$ blood

ⓔ **1/1 mark awarded** The answers must include the units to gain the mark.

(ii) The fetus gains oxygen across the placenta. The pO_2 in the placenta is low so maternal oxyhaemoglobin releases oxygen. At $4\,kPa$ fetal haemoglobin has a *higher* affinity for oxygen, so this maintains the diffusion gradient between maternal and fetal blood.

ⓔ **3/3 marks awarded** Practise taking readings from these graphs by using a ruler to read *vertically* at different values of pO_2. The vertical axis is equivalent to the *affinity* of haemoglobin for oxygen.

(iii) Oxygen would not be released readily enough because the affinity of fetal haemoglobin would be too high. This means it would be difficult for them to gain enough oxygen for the high metabolic rate of children and adults. If a woman with this condition becomes pregnant then the fetus will gain very little oxygen.

ⓔ **2/2 marks awarded** Link the answer to physiology, giving the consequences of a limited supply of oxygen to tissues.

(d) The zinc ion is the prosthetic group in carbonic anhydrase, which catalyses the reaction between water and carbon dioxide as blood flows through capillaries in respiring tissues to form carbonic anhydrase. The ion forms part of the active site of the enzyme. Carbonic acid dissociates into hydrogen ions and hydrogencarbonate ions. The latter ions diffuse out of red blood cells. Chloride ions diffuse into the cells to maintain electrochemical neutrality in the chloride shift.

🅔 **5/5 marks awarded** At A-level you can expect to use ideas and factual knowledge from two or more modules; here we have Modules 2 (ions and enzymes) and 3 (transport of carbon dioxide).

Question 8

Researchers investigated the diversity of macro (large) invertebrates in soils in a tropical area of Zambia. They took soil samples from three areas:

- natural woodland
- monoculture of maize
- maize grown with various tree species

Samples were taken in December during the wet season and in July during the dry season. Soil to a depth of 250 mm was sampled. The samples were 3–5 m apart. The stratified sampling technique was used within each area. The macro invertebrates were identified as far as possible and counted. The results are shown in Figure 1.

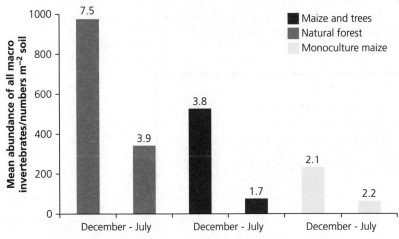

Figure 1 The numbers across the top represent the mean richness per sample from each of the six areas

(a) Explain:
   (i) why the researchers used a stratified sampling technique in areas 1 and 3 (2 marks)
   (ii) why the mean richness per sample is not expressed as species richness (1 mark)

🅔 Take care in part (a)(i). This does not, as you might think, mean sampling at different heights or depths. The clue to part (a)(ii) is in the stem of the question. Use your knowledge of the different taxonomic ranks to answer this with some technical vocabulary.

(b) Use the results in the bar chart to compare the abundance and diversity of macro invertebrates in the soil samples from the different areas in the wet and dry seasons.

(6 marks)

ⓔ Spend time analysing the results and making notes before writing an answer. Take data from the bar chart and use data comparatively, using words like 'higher' and 'lower' and quoting figures in evidence. This is another skill you will have developed by the time you take A-level papers.

The table shows the abundance of six of the groups of macro invertebrates in the soil samples.

	Macro invertebrate groups	Natural forest		Maize and trees		Monoculture of maize	
		Dec	July	Dec	July	Dec	July
Mean abundance/ numbers m^{-2}	Beetles (d, c1 and c2)*	112.0	44.4	24.4	3.1	14.2	10.7
	Earthworms (d)	48.0	0.9	8.4	0.0	7.1	0.0
	Ants (d)	246.0	142.0	49.5	22.2	8.9	24.9
	Termites (d)	396.0	113.8	392.7	32.9	197.3	7.1
	Centipedes (c2)	24.0	4.4	7.3	0.4	1.8	0.0
	Millipedes (d and c1)	72.0	17.8	17.1	4.4	0.0	0.0

* c1 = primary consumer; c2 = secondary consumer; d = detritivore

(c) Use the results in the table to comment on the effect of growing maize with trees and as a monoculture on the abundance of macro invertebrates.

(6 marks)

ⓔ The question says 'comment...', so you can describe the results, explain them and evaluate them. This question is set in the context of the ecology topics in Module 6.

ⓔ 6/6 marks awarded Notice the use of data taken from the table to support the different ideas. You need to identify the ideas first and then choose the most appropriate data.

(c) Examples of descriptions of the data in Table 2:

- The abundance of all groups shown in the table is less in the monoculture plots in both months, with the exception of ants in July.
- In December, the abundance of macro invertebrates is significantly lower in maize grown with trees and monoculture of maize than in the natural forest.
- The abundance is lower in monocultures of maize in July than in the maize with trees.

Examples of possible explanations:

- Monoculture has very little food available to soil invertebrates, because there is no leaf fall from forest trees.
- The crop debris is cleared away and not left on the soil for detritivores to eat.
- The soil is exposed to the atmosphere so there is no protection from predators and changes in temperature.
- Soil is disturbed by cultivation, so large invertebrates move away or are predated.

Examples of evaluation:

- Only large invertebrates were sampled (soils support many smaller invertebrates).
- There are no results for decomposers (fungi and bacteria).
- There is no indication of the number of samples per plot.
- There is no indication of the repeatability of the results.

ⓔ 6/6 marks awarded These are only some of the comments that could be made for the evaluation. It is always a good idea to put yourself in the place of the experimenters and think about what they did well and what they could have done better.

Answers to multiple-choice questions

Question	AS	A2
1	D	B
2	A	D
3	A	D
4	D	C
5	B	B

Knowledge check answers

1 diameter = 5.30 mm; radius (r) = 2.65
length (l) = 21.00
volume of a cylinder = $\pi r^2 l$
= $3.142 \times (2.65)^2 \times 21.00 = 463.36 \, \text{mm}^3$
surface area of a cylinder = $2\pi r^2 + 2\pi r l$
= $(2 \times 3.142 \times (2.65)^2) + (2 \times 3.142 \times 2.65 \times 21.00)$
= $44.13 + 349.70 = 393.83 \, \text{mm}^2$
SA:V ratio = $\dfrac{SA}{V} = \dfrac{393.83}{463.36} = 0.85{:}1$
You should know how to calculate the surface areas and volumes of rectangular, cylindrical and spherical shapes. Many cells have shapes like these.

2 mouth/nose → throat → trachea → bronchus → bronchiole → alveolus

3 There are many of them so they have a large surface area; the squamous epithelium is thin so there is a short diffusion distance; there are many capillaries around each alveolus, allowing for uptake of much oxygen.

4 a Ventilation is the movement of air into and out of the lungs. Gas exchange is the diffusion of carbon dioxide and oxygen between air and the body.

   b Respiration is the series of chemical reactions that occur in cells to release energy from biological molecules so that it is available for energy-requiring processes. Breathing refers to the methods by which animals ventilate their gas exchange surface.

5 During shallow breathing the ribcage does not move, but the abdomen moves out when you breathe in and moves inwards when you breathe out. During deep breathing the ribcage moves up and out when breathing in, and down and inwards when breathing out. The movement of the abdomen is the same as in shallow breathing.

6 insects — tracheoles; bony fish — secondary lamellae in gills; mammals — alveoli

7 In an open system the blood bathes the cells directly; it does not flow through blood vessels. In a closed system blood flows inside vessels.

8 In a single system blood flows through the heart once during a complete circuit of the body. In a double system blood flows through the heart twice in one complete circuit.

9 Blood in the pulmonary circulation flows from the heart to the lungs and back; in the systemic circulation blood flows from the heart to all the other organs and back to the heart.

10 stroke volume = $\dfrac{\text{cardiac output}}{\text{heart rate}}$
$\dfrac{7.50}{90} = 0.083 \, \text{dm}^3$ or $83 \, \text{cm}^3$

11 (vena cava) → right atrium → right ventricle → pulmonary artery → capillaries in the lung → pulmonary vein → left atrium → left ventricle (aorta)
This is *not* the cardiac cycle — this is the pathway taken by blood through the heart and through the pulmonary circulation.

12 The SAN initiates the heart beat by emitting waves of depolarisation or excitation. The AVN relays these waves to the Purkyne tissue, which conducts them to the base of the ventricles.

13 The graph should have the same axes; blood pressure changes in the right atrium are the same as in the left; the maximum values for the right ventricle and the pulmonary artery are significantly lower than the left ventricle and the aorta. The highest pressure in the right ventricle should not be more than 4.0 kPa.

14 There are two separate circulations for oxygenated and deoxygenated blood. Blood can be pumped to respiring tissues at high pressure with efficient delivery of oxygen. Blood is pumped to the lungs at a lower pressure.

15 Examples include: aorta, hepatic artery and vein (to and from liver), renal artery and vein (kidney), coronary artery (heart), femoral artery and vein (leg), vena cava.

16 The curve for fetal haemoglobin should be the same shape as that for adult haemoglobin and a little to the left. This is because fetal haemoglobin has a higher affinity for oxygen, so at each partial pressure of oxygen it has a higher percentage saturation. You can see this if you put a ruler vertically on your graph against any of the values for partial pressure of oxygen.

17

	Transpiration	Translocation
Sources	Root	Leaves
Sinks	Leaves, stems, flowers and fruits	Storage organs, stems, flowers and fruits

At the beginning of the growing season storage organs (e.g. potato tubers and carrot roots) act as sources for sucrose and amino acids and the growing leaves are sinks.

18

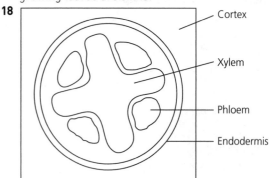

19 (soil) → root hair cell → cortex (apoplast) → endodermis (symplast) → xylem vessel → leaf → mesophyll cell (evaporation) → air space → stoma → (atmosphere)

20 Water is a raw material for photosynthesis. Water is absorbed by cells to maintain their turgidity.

21 Active immunity involves an immune response; production of antibodies within the body; formation of memory cells. None of these things happen in passive immunity. The first response in active immunity is slower than passive immunity; active immunity provides long-term protection, passive immunity is temporary.

22 Memory cells enable a fast response during secondary immune responses to an antigen. This prevents an infection by the same antigen causing disease.

23 Secondary is faster and involves production of a higher concentration of antibodies than during the primary response.

24 Influenza antigens change from year to year as a result of mutation and cross-breeding between strains of influenza viruses. The body does not recognise these new antigens and so immunity to previous infections or vaccinations does not provide protection against new strains.

25 An antibiotic is a medicinal drug used to treat bacterial diseases (they are not effective against viruses). An antibody is a protein secreted by activated B lymphocytes (plasma cells) on stimulation by an antigen.

26 Species richness is the number of species present in an area. Species evenness is the abundance of the different species in an area.

27 16 of the genes were polymorphic, i.e. 16 had two or more alleles with a frequency >5% (or >1%) in the population studied.

28 Species richness and evenness; genetic diversity within species; ecosystem diversity

29 Taxonomic rank: a group used in classification, e.g. kingdom, class and species.

Hierarchy: a system of organisation in which each group is subdivided into further groups; in classification a kingdom is divided into phyla, each phylum is divided into classes, etc.

Binomial system: system for naming organisms. The name of each species consists of the name of its genus (generic name) and its specific name (also known as its specific epithet).

Phylogeny: the evolutionary history of a taxon, for example a species.

30 A feature (physiological, behavioural, etc.) that helps an organism 'fit' into its habitat, breed and pass on its alleles.

Index

Index

Index